Soil Genesis
and Classification

Soil Genesis and Classification

Dr. Parmeshwar Singh

Editor

KOROS PRESS LIMITED
London, UK

Soil Genesis and Classification

© 2012

Printed in 2017 for Sale in the Indian Subcontinent

Published by
Koros Press Limited
3 The Pines, Rubery B45 9FF, Rednal,
Birmingham, United Kingdom

Tel.: +44-7826-930152
Email: info@korospress.com
www.korospress.com

ISBN: 978-1-78163-115-7

Editor: Dr. Parmeshwar Singh

Printed in UK

British Library Cataloguing in Publication Data
A CIP record for this book is available from the British Library

10 9 8 7 6 5 4 3 2 1

Exclusively distributed by CBS Publishers & Distributors Pvt. Ltd.

Sales & Distribution Rights only for India, Pakistan, Bangladesh, Sri Lanka, Nepal and Bhutan.This book is not to be sold outside these territories.

Contents

Analysis: Key to Nutrient Management Planning • What is
Measured? • Factors Affecting Soil Analysis • Sampling
• Interpretation of Soil pH • Other Factors Affecting
Interpretation • Changes in Soil Nutrient Status • Improving
Soils with Low Nutrient Status • Treatment of High Fertility
Soils • Analysis and Collection of Soil Samples

5. Soil pH and Electrical Conductivity   133

Soil pH and its Uses • Nitrogen Fixing Microbes • Liming of
Soils • Electrical Conductivity and its Uses • Interpretation of
Electrical Conductivity of Irrigation Water • Interpretation of
Electrical Conductivity of Soils • Laboratory Procedures • Soil
Electrical Conductivity Variability • Methods • Results and
Discussion

6. Soil Moisture Measurement Instrumentation   151

Introduction • Instruments Available for Objective Measurement
of Soil Moisture Content • Electrical Resistance (Gypsum)Blocks
• Some Diseases of the Soil • The Formation of Alkali Land
• The Maintenance of Soil Fertility in Great Britain • The
Roman Occupation • The Saxon Conquest • The Open-Field
System • The Depreciation of Soil Fertility • The Low Yield
of Wheat • The Black Death • The Industrial Revolution and
Soil Fertility • The Great Depression of 1879 • The Second
World War

7. Examination and Description of Soils   190

Introduction • Some General Terms Used in Describing Soils
• Ground Surface Cover • Material Produced by Weathering
of Rock in Place • Material Moved and Deposited by Wind
• Material Moved and Deposited by Glacial Processes • Erosion
• Water Erosion • Soil Temperature • Designations for Horizons
and other Layers • Cyclic and Intermittent Horizons and Layers
• Boundaries of Horizons and Layers • Soil Structure
• Concentrations • Consistence • Animals

Preface

Soils are made up of four components: minerals, air, water, and organic matter. In most soils minerals represent around 45% of the total volume, water and air about 25% each, and organic matter from 2% to 5%. The mineral portion consists of three distinct particle sizes classified as sand, silt or clay. Sand is the largest size particle that can be considered soil. Sand is largely the mineral quartz, though other minerals are also present. Since quartz contains no plant nutrients, sand is the lowest contributor to soil fertility of the three soil particle sizes. Furthermore, sand cannot hold nutrients–they leach out easily with rainfall. That is why sandy soils are not as productive as loams and need to be spoon-fed fertilizer. Silt particles are much smaller than sand but, like sand, silt is mostly quartz. The smallest of all the soil particles is clay. Clays are quite different from sand or silt and contain appreciable amounts of plant nutrients. Clay has a large surface area resulting from the plate-like shape of the individual particles. The textural designation of a soil is derived from the relative portions of sand, silt, and clay. A sandy loam, for example, has much more sand and much less clay than does a clay loam. A loam soil is a mixture of sand, silt and clay. Most soils are some type of loam.

Another soil characteristic–soil structure–is different from soil texture. Structure refers to the combination or "aggregation" of sand, silt and clay particles into larger secondary clusters. If you grab a handful of soil, good structure is apparent when the sand, silt, and clay particles are aggregated into granules or crumbs. Both texture and structure determine pore space for air and water circulation, erosion resistance, looseness, ease of tillage, and root penetration. However, while texture is an innate property of the native soil and does not change with agricultural activities, structure can be improved or destroyed readily through our choice and timing of farm practices. The organic soil component contains all the living creatures in the soil and the dead ones in various stages of decomposition. An acre of living soil can contain 900 pounds of earthworms, 2400 pounds of fungi, 1500 pounds of bacteria, 133 pounds of protozoa, 890 pounds of arthropods and algae, and even small mammals in some cases. In fact, the soil could be viewed as a living entity, rather than an inert body.

The soil's organic matter also contains dead organisms, plant matter and other organic materials in various phases of decomposition. Humus, the dark-coloured organic material in the final stages of decomposition, is relatively stable. Both organic matter and humus serve as a reservoir of plant nutrients; they also help to build soil structure and provide other benefits. The type of healthy living soil required to support humans now and far into the future will be balanced in nutrients and high in humus with a high diversity of soil organisms. It will produce healthy plants with minimal weed, disease and insect pressure. To accomplish this we work with the natural processes and optimize their functions to sustain our farms.

This book is designed to be a work in progress. The information provides a starting place from which improved sufficiency ranges can be developed. Revisions will be published as additional information becomes available.

—Editor

Chapter 1

Introduction

Soils

Sampling: Soil sampling is a particularly difficult task when attempting to get a representative sample. Normally a 500-gram sample is submitted to the laboratory for analysis. This 500-gram sample may represent 10 or more acres. If the area covered by the sample is not uniform, the chemical analysis may not accurately reflect the nutrient status of specific sites. Factors that need to be considered when sampling soil include the depth and time of sampling. Proper sampling depth is affected by the crop being grown, past cropping, depth of plowing and also the nutrient of interest. Subsoil samples are important for most crops. Standard sampling times should be used due to the difficulty in comparing samples taken at different times. The fertility level of a field will vary over the course of the year and interpreting results for samples taken at different times of the year will be very difficult. Sampling between crops will give more consistent results.

When sampling soils, the area should be subdivided into as homogeneous sections as possible. Between 10 and 20 sub-samples should be composited from each area. Sub-samples should be small enough that the composite sample will be of a size that can be completely processed for analysis. The depth of the sampling is determined by the crop, the elements of interest and existing knowledge about the soil profile. Samples for cultivated crops are taken from the plow layer. Pasture and sod crop soil samples should be collected from the top four inches. Samples for nitrate, soluble salts and available micronutrients should be taken at the root depth.

Drying Recommendations: Once samples have been collected, they must be processed promptly to prevent any changes that might affect the analysis. Break up large chunks of soil and spread out to air dry where the sample will not be contaminated, particularly by fertilizer dust. The sample may also be placed in a forced air oven set between 35°C and 55°C.

Grinding Recommendations: Soil should be crushed in a soil pulverizer to pass through a 20 mesh screen. Large clumps of hard soil should be pounded into smaller clumps first. Soil pulverizing time is normally one to four minutes per sample.

Exceptions:

- Analysis of Carbon, Nitrogen (including Total Kjedahl Nitrogen), and total elements (Arsenic, Cadmium, Copper, Chromium, Iron, Lead, Manganese, Molybdenum, Nickel, Phosphorus, Selenium, Zinc) requires that the sample be ground to a powder fine enough to pass through a 60 mesh screen.
- Soils with a high percentage of sand or organic material may require grinding to pass through a 60 mesh screen to achieve sample homogeneity.

Plants

Sampling: Sampling plays a critical role in plant analysis. When analyzing the nutrient status of plants, it is essential to select the plant part for chemical analysis that reflects the status of the particular element of interest. Four samplings during a growing season are usually sufficient to characterize seasonal nutritional patterns.

One sampling should be early in the growing season, two in mid-season and the last one just prior to harvest. Four samples should be collected from each field or management unit. Each sample should contain material from at least 20 plants to ensure adequate, representative material for analytical testing. Separate samples should be taken from areas that appear different from the rest of the field.

A young mature leaf is generally selected for analysis. The sample can be subdivided into blade and petiole. The status of Cl, NO_3-N, NH_4-N, extractable K and P, in the form of PO_4-P (2% acetic acid) are generally determined through analysis of the petiole. Blades are used when evaluating the status of K, Ca, Mg, Na, Fe, Mn, Zn, Cu, B, Mo, SO_4-S and total-N in plants. For diagnostic purposes, only leaves that have recently developed symptoms should be collected for chemical analysis.

Drying Recommendations: After collection, plant material should be washed to remove any residual soil or dust. Fresh samples or those suspected to be moist should be placed into paper bags (with adequate room for air movement within the bag) and dried in a forced air oven at 55-60°C. In general, adequate drying time is approximately 12 hours or until the material snaps or breaks easily. All samples, except freeze drying material, should be turned every 24 hours.

Exceptions:

- Range samples take 24 hours or more to dry
- Power plant ash samples take three days to dry
- Solid pieces of wood take 24 hours to dry
- Rice soils take two days to dry
- Any material larger than gallon size will take one or more days to dry
- Fruits and vegetables should be freeze dried and pureed (weigh the samples before and after the freeze drying process).

These time frames are based on use of a large, forced-air oven. Times need to be adjusted accordingly for other types/sizes of ovens.

Grinding Recommendations: Most plant and feed samples should be ground to pass through a 40 mesh screen. Large wood samples must be splintered into smaller pieces before grinding. Average grinding time is one minute per sample if less than 10 grams or three to four minutes for samples over 10 grams.

Exceptions:

- Grape blades-use a 20 mesh screen due to tricomb separation from leaf
- Sugar beet petioles-use a 20 mesh screen due to high sugar content
- Walnut leaf-use a 20 mesh screen due to fibers in veins
- Corn stalks-use a 10 mesh screen due to high sugar content
- Freeze dried samples-do not need to be ground as they are easily crushed manually.

Water

Sampling: Samples must be representative. Samples should be collected in clean, plastic bottles that have been rinsed three times prior to use. Well-water samples should be collected after pumping for at least 30 minutes. Sampling from distribution systems should be done after the lines have been flushed sufficiently to ensure that the sample is representative of the supply.

If NO_3-N, NH_4-N or PO_4-P are elements of interest the samples should be frozen or kept below 40° F. Samples collected for alkalinity, conductivity, phosphate, sulfate, turbidity or solids should also be refrigerated until analysis can be completed. Due to the problems of absorption or precipitation, if micronutrients, metals or salts are of interest, the sample, or a sub-sample, should be acidified to pH <2 as soon as possible after collection. Freezing is a good choice for water sample preservation in many cases. Samples high in carbonates should not be frozen.

Please take the following precautions when sending frozen water samples:

- Always leave at least 10% of the container unfilled to accommodate the expansion of the water when frozen.
- Do not place sample containers directly onto dry ice. Place some insulation between the dry ice and the sample containers.
- Do not use glass bottles the sheer bulk of the sample becomes significant. Soil samples of over 50 mg are very difficult to analyse due to rapid ash buildup in the furnace.

Combustion capsule formation is critical for successful analytical runs. Please refer to the detailed instructions for further information.

Setting the Standard in Soil Sampling

New standards when sampling for volatiles in soil will demand new processes and new equipment according to Vincent Van Walt.

If you are involved in sampling for volatile compounds in soils – things have changed with new regulations and standards coming into force in the USA (USA-EPA Guideline 5035a_r1) and with similar in preparation in Europe through the ISO. In summary these new standards recommend that less soil (16ml ~ 25g) is taken for laboratory analysis and that any soil to be investigated should be sampled with a limited amount of disruption during the extraction process in order to minimise VOC losses prior to sample preparation and determination in the laboratory. This usually means collecting the soil sample with a single transfer to an air tight container that will be used for storage and preparation for analysis. In addition, acceptable sampling devices, procedures, preservatives and techniques are required to be recorded and observed.

Previously samples of soils for the analysis of very volatile components such as benzene, toluene and chlorinated hydrocarbons were normally obtained directly from a borehole. The stainless steel sample tubes containing the soil were closed completely and cooled for further transport to the laboratory for analysis. The new regulations recommend that a small, accurately measured sub-sample be taken above ground from a larger sampler or auger.

For environmentalists in the field, this new closed-system purge-and-trap process for the analysis of volatile compounds in solid materials like soils, sediments, and solid waste requires different equipment. For a low soil method a hermetically-sealed sample vial is required, the seal of which is never broken from the time of sampling to the time of analysis and since the sample is never exposed to the atmosphere after sampling, the losses of VOCs during sample transport, handling, and analysis should be minimised. The applicable concentration range of the low soil method is dependent on the determinative method, matrix, and compound. However, it will generally fall in the 0.5 to 200ìg/kg range.

As with any preparative method for volatiles, samples should be screened to avoid contamination of the purge-and-trap system by samples that may contain very high concentrations of purgeable materials. In addition, because the sealed sample container cannot be opened to remove a sample without compromising the integrity of the sample, multiple samples should be collected to allow for screening and reanalysis.

However, not all closed-system purge-and-trap equipment employed for low concentration samples is appropriate for soil samples preserved in the field with methanol. Precautions should be taken when preserving a soil by this method since certain compounds within the olefins, ketones, esters, ethers, and sulfides classes may react under low pH conditions and possibly not be representative of the material sampled. Additionally, acidification of solid wastes may evolve toxic gases that may be harmful to field and laboratory personnel. It is therefore recommended that when collecting wastes of unknown composition, preliminary screening and characterisation of potential sample contents should be performed prior to use of acidification as a means to chemically preserve samples designated for determinative analyses.

After a fresh surface of the solid material to be analysed is exposed to the atmosphere, the sub-sample collection process should be completed as quickly as possible in order to minimise the loss of VOCs due to volatilisation. Removing the sub-sample from a material should be done with the least amount of disruption (disaggregation) as possible. Additionally, rough trimming of the sampling location's surface layers should be considered if the material may have already lost VOCs (been exposed for more than a couple of minutes) or if it may be contaminated by other waste, different soil strata, or vegetation. Removal of surface layers can be accomplished by scraping the surface using a clean spatula, scoop, knife, or shovel.

Sub-sampling should be carried out using an appropriate device. The new standards recommend a smaller coring device that will help to maintain the sample structure during collection and transfer as do their larger counterparts used to retrieve subsurface materials. When inserting a clean coring tool into a fresh surface for sample collection, air should not be trapped behind the sample. If air is trapped, it could either pass through the sampled material causing VOCs to be lost or cause the sample to be pushed prematurely from the coring tool.

After an undisturbed sample has been obtained by pushing the barrel of the coring tool into a freshly exposed surface and then removing the corer once filled, the exterior of the barrel should be quickly wiped with a clean disposable towel. The next step varies, depending on whether the coring device is used for sample storage and transfer or solely for transfer. If the coring tool is used as a storage container, cap the open end after ensuring that the sealing surfaces are cleaned. If the device is to be solely used for collection and not for storage, immediately extrude the sample into a VOA vial or bottle. The volume of material collected should not cause excessive stress on the coring tool during intrusion into the material, or be so large that the sample easily falls apart during extrusion.

Obtaining and transferring a sample should be done rapidly (<10 seconds) to reduce volatilisation losses. If the vial or bottle contains methanol or another liquid, it should be held at an angle when extruding the sample into the container to minimise splashing. Just before capping, a visual inspection of the lip and threads of the sample vessel should be made, and any foreign debris should be removed with a clean towel, allowing an airtight seal to form.

Following the new standards, a site-specific Sampling and Analysis Plan should clearly list the required sample collection equipment necessary to ensure that the loss of volatile constituents is minimised during the sample collection process. As with all environmental sampling applications, the analytical data usability and representativeness will be affected by improper sample collection techniques.

Sampling personnel will be responsible for ensuring that VOA vials are sealed properly using a septum of sufficient thickness without any punctures. The improper vial sealing (i.e. due to excess sample retained on the vial threads) and tightening of caps are the primary factors in the loss of volatiles due to sample collection activities. Care should also be exercised in the selection of approved pre-cleaned and certified VOA vials absent of burrs on the glass.

Procedures should be in place for the selection and appropriate use of sample collection devices (i.e. bailer, coring tool, etc.) along with the required decontamination measures. It is also recommended to store a control sample when collecting volatile samples in order to assess possible field induced contamination.

Whichever method is most appropriate for the volatile and soil conditions to be sampled the equipment chosen should be easy to decontaminate so a high grade stainless steel is ideal. Aluminium lined seals will prevent sorption and diffusion and, the mechanism of the corer should aim to eliminate headspace and therefore loss of VOCs from the soil.

The new standards will improve the quality of results obtained when sampling for volatiles in soil however the problem of obtaining the sample should not be understated. The problem lays in the field not the laboratory so the soil sampler used, the time of exposure prior to sub-sampling, wind conditions, temperature and the type of volatile being tested for will all impact on the potential losses.

There is no doubt with the new standards preservation of the volatiles is optimised and freezing of the sample mitigates volatilisation but it is still the first steps in the process that remain the Achilles heel-but who am I to argue with the esteemed authors of guidelines or standards?

The Living Soil: Texture and Structure

Soils are made up of four components: minerals, air, water, and organic matter. In most soils minerals represent around 45% of the total volume, water and air about 25% each, and organic matter from 2% to 5%. The mineral portion consists of three distinct particle sizes classified as sand, silt or clay. Sand is the largest size particle that can be considered soil. Sand is largely the mineral quartz, though other minerals are also present. Since quartz contains no plant nutrients, sand is the lowest contributor to soil fertility of the three soil particle sizes. Furthermore, sand cannot hold nutrients–they leach out easily with rainfall. That is why sandy soils are not as productive as loams and need to be spoon-fed fertilizer. Silt particles are much smaller than sand but, like sand, silt is mostly quartz.

The smallest of all the soil particles is clay. Clays are quite different from sand or silt and contain appreciable amounts of plant nutrients. Clay has a large surface area resulting from the plate-like shape of the individual particles. The textural designation of a soil is derived from the relative portions of sand, silt, and clay. A sandy loam, for example, has much more sand and much less clay than does a clay loam. A loam soil is a mixture of sand, silt and clay. Most soils are some type of loam. They are more accurately described by the words the preface the word loam, such as: sandy loam or clay loam. The texture designations are found in Table.

Table 1 : Soil textures

Texture Designation	
Coarse Textured	Fine
Textured	Sand
Loamy sand	
Sandy loam	
Fine sandy loam	
Loam	
Silty loam	
Silt	
Silty clay loam	
Clay loam	
Clay	

Another soil characteristic–soil structure–is different from soil texture. Structure refers to the combination or "aggregation" of sand, silt and clay particles into larger secondary clusters. If you grab a handful of soil, good structure is apparent when the sand, silt, and clay particles are aggregated into granules or crumbs.

Both texture and structure determine pore space for air and water circulation, erosion resistance, looseness, ease of tillage, and root penetration. However, while texture is an innate property of the native soil and does not change with agricultural activities, structure can be improved or destroyed readily through our choice and timing of farm practices. The organic soil component contains all the living creatures in the soil and the dead ones in various stages of decomposition. An acre of living soil can contain 900 pounds of earthworms, 2400 pounds of fungi, 1500 pounds of bacteria, 133 pounds of protozoa, 890 pounds of arthropods and algae, and even small mammals in some cases. In fact, the soil could be viewed as a living entity, rather than an inert body.

The soil's organic matter also contains dead organisms, plant matter and other organic materials in various phases of decomposition. Humus, the dark-coloured organic material in the final stages of decomposition, is relatively stable. Both organic matter and humus serve as a reservoir of plant nutrients; they also help to build soil structure and provide other benefits.

The type of healthy living soil required to support humans now and far into the future will be balanced in nutrients and high in humus with a high diversity of soil organisms. It will produce healthy plants with minimal weed, disease and insect pressure. To accomplish this we work with the natural processes and optimize their functions to sustain our farms.

The Living Soil: Importance of Soil Organisms

Figure 1: The soil is teaming with organisms which cycle nutrients from soil to plant and back again

If you look out at a landscape you might wonder how native prairies and forests function in the complete absence of tillage and fertilizers? These soils are tilled by soil organisms, not by machinery. They are fertilized too, but the fertility is used again and again and never leaves the site. Native soils are covered with a layer of plant litter and/or growing plants throughout the year. Beneath the surface litter layer, a rich complexity of soil organisms decompose plant residue and dead roots, then release their stored nutrients slowly over time.

In fact, topsoil is the most biologically diverse part of the earth. Soil-dwelling organisms release bound-up minerals converting them into plant-available forms that are then taken up by the plants growing on the site. The organisms recycle nutrients again and again from the death and decay of each new generation of plants growing on the site.

There are many different types of creatures that live on or in the soil. Each has a role to play. These organisms will work for the farmer's benefit if we simply manage for their survival. Consequently we may refer to them as soil livestock. While there is a great variety of organisms that contribute to soil fertility, earthworms, arthropods, and the various microorganisms merit particular attention.

Earthworms: Earthworm burrows enhance water infiltration and soil aeration. Earthworm tunnelling can increase the rate of water entry into the ground 4 to 10 times higher than fields that lack worm tunnels. This reduces water runoff, recharges groundwater, and helps store more soil water for dry spells. Vertical earthworm burrows pipe air deeper into the soil, stimulating microbial nutrient cycling at those deeper levels. Tillage done by earthworms can replace some expensive tillage work done by machinery.

Worms eat dead plant material left on top of the soil and redistribute the organic matter and nutrients throughout the topsoil layer. Nutrient-rich organic compounds line the tunnels that may remain in place for years if not disturbed. During droughts these tunnels allow for deep plant root penetration into subsoil regions of higher moisture content. In addition to organic matter, worms also consume soil and soil microbes as they move through the soil. The soil clusters they expel from their digestive tracts is known as a worm cast or casting. Each worm cast is separate from other casts and ranges in size from that of a mustard seed to a sorghum seed depending on the size of the worm. The soluble nutrient content of worm casts is considerably higher than those of the original soil (see Table next page). A good population of earthworms can process 20,000 pounds of topsoil per year, with turnover rates as high as 200 tons per acre having been reported in some exceptional cases.

Earthworms also secrete a plant growth stimulant. Reported increases in plant growth due to earthworm activity may be attributed to this substance, not just to improved soil quality.

Table 2: Nutrients in worm casts compared to the surrounding soil from which they came.

Nutrient	Worm casts	Soil
	Lbs/ac	Lbs/ac
Carbon	171,000	78,500
Nitrogen	10,720	7000
Phosphorus	280	40
Potassium	900	140

Earthworms thrive where there is no-tillage–generally, the less tillage, the better, and the shallower the tillage, the better. Worm numbers can be reduced by as much as 90% by deep and frequent tillage. Tillage reduces earthworm populations by drying the soil, burying the plant residue they feed on, and making the soil easier to freeze. Tillage also destroys their vertical burrows and can kill and cut up the worms themselves. Emergence times for young worms are spring and fall–their most active periods just when most farmers are interested in tillage. Worms are dormant in the hot part of the summer and the cold of winter. Table below shows the effect of tillage and cropping practices on earthworm numbers.

Table 3: Effect of crop management on earthworm populations.

Crop	Management	Worms/foot2
Corn	Plow	1
Corn	No-till	2
Soybean	Plow	6
Soybean	No-till	14
Bluegrass/clover	—	39
Dairy pasture	—	33

As a rule, earthworm numbers can be increased by reducing or eliminating tillage (especially fall tillage), never using the moldboard plow, reducing residue particle size (using a straw chopper on the combine), adding animal manure, and growing green manure crops. It is beneficial to leave as much surface residue as possible year round. Cropping systems that typically have the most earthworms are in descending order: perennial cool-season grass grazed rotationally, then warm-season perennial grass grazed rotationally, then annual croplands using no-till. Ridge-till and strip tillage will generally have more earthworms than clean tillage involving plowing and disking.

Earthworms prefer a near neutral soil pH, moist soil conditions, and plenty of plant residue on the soil surface. They are sensitive to certain pesticides and some incorporated fertilizers. Carbamate insecticides including Furadan, Sevin, and Temik, are harmful to earthworms, notes worm biologist Clive Edwards of Ohio State University. Some insecticides in the organophosphate family are mildly toxic to earthworms while synthetic pyrethroids are harmless to them. Most herbicides have little effect on worms except for the triazines, such as Atrazine, which are moderately toxic. Also, anhydrous ammonia kills earthworms in the injection zone.

Arthropods: In addition to earthworms, there are many other species of soil organisms that can be seen by the naked eye. Among them are sowbugs, millipedes, centipedes, slugs, snails and springtails. These are the primary decomposers.

Their role is to eat and shred the large particles of plant and animal residues. Some bury residue, bringing it into contact with other soil organisms that further decompose it. Some members of this group prey on smaller soil organisms. The springtails are a small insect, which eat mostly fungi. Their waste is rich in plant nutrients that are released after other fungi and bacteria decompose it. Also of interest are dung beetles, which play a valuable role in recycling manure and reducing livestock intestinal parasites and flies.

Bacteria: Most numerous among soil organisms are the bacteria; every gram of soil contains at least a million of these tiny one-celled organisms. There are many different species of bacteria, each with its own role in the soil environment.

One of the major benefits bacteria provide for plants is in helping them take up nutrients. Some species release nitrogen, sulfur, phosphorus, and trace elements from organic matter. Others break down soil minerals and release potassium, phosphorus, magnesium, calcium and iron. Still other species make and release natural plant growth hormones, which stimulate root growth.

A few species of bacteria fix nitrogen in the roots of legumes while others fix nitrogen independently of plant association. Bacteria are responsible for converting nitrogen from ammonium to nitrate and back again depending on certain soil conditions. Other benefits to plants provided by various species of bacteria include increasing the solubility of nutrients, improving soil structure, fighting root diseases, and detoxifying soil.

Fungi: Fungi come in many different species, sizes and shapes in soil. Some species appear as thread-like colonies, while others are one-celled yeasts. Slime molds and mushrooms are also fungi. Many fungi aid plants by breaking down organic matter or by releasing nutrients from soil minerals. Fungi are generally early to colonize larger pieces of organic matter and begin the decomposition process. Some fungi produce plant hormones, while others produce antibiotics including penicillin. There are even species of fungi that trap harmful plant-parasitic nematodes.

The mycorrhizae (my-cor-ry-'zee) group of fungi lives either on or in plant roots and act to extend the reach of root hairs into the soil. Mycorrhizae increase the uptake of water and nutrients especially in less fertile soils. Roots colonized by mycorrihizae are less likely to be penetrated by root-feeding nematodes since the pest cannot pierce the thick fungal network. Mycorrhizae also produce hormones and antibiotics, which enhance root growth and provide disease suppression. The fungi benefit from plant association by taking nutrients and carbohydrates from the plant roots they live in.

Actinomycetes: Actinomycetes are thread-like bacteria that look like fungi. While not as numerous as bacteria, they also perform vital roles in the soil. Like the bacteria, they help decompose organic matter into humus, releasing nutrients. They also produce antibiotics to fight diseases of roots. These same antibiotics are used to treat human diseases. Actinomycetes are responsible for the sweet, earthy smell of biologically active soil noticed whenever a field is tilled.

Algae: Many different species of algae also live in the upper half-inch of the soil. Unlike most other soil organisms, algae actually produce their own food through photosynthesis. They appear as a greenish film on the soil surface following a good rain. Algae improve soil structure by producing slimy substances that glue soil together into water-stable aggregates. Some species of algae (the blue-greens) can fix their own nitrogen, some of which is later released to plant roots.

Protozoa: Protozoa are free-living microorganisms that crawl or swim in the water between soil particles. Many soil protozoa are predatory, eating other microbes. One of the most common is an amoeba that eats bacteria. By eating and digesting bacteria, protozoa speed up the cycling of nitrogen from the bacteria, making it more available to plants.

Nematodes: While nematodes are abundant in most soils, only a few species are harmful to plants. The harmless species eat decaying plant litter, bacteria, fungi, algae, protozoa and other nematodes. Like other soil predators, nematodes speed the rate of nutrient cycling.

All these organisms–from the tiny bacteria up to the large earthworms and insects–interact with one another in a multitude of ways in a whole soil ecosystem. Organisms not directly involved in decomposing plant wastes may feed on each other or each other's waste products or the other substances they release. Among the other substances released by the various microbes are vitamins, amino acids, sugars, antibiotics, gums, and waxes.

Roots can also release various substances into the soil that stimulate soil microbes. These substances serve as food for select organisms. Some scientists and practitioners theorize that plants use this means to stimulate the specific population of microorganisms capable of releasing or otherwise producing the kind of nutrition needed by the plants. Research on biological life in the soil has determined that there are ideal ratios for certain key soil organisms in highly productive soils (soil foodweb).

The Soil Foodweb lab, located in Oregon, tests soils and makes fertility recommendations that are based on this understanding. Their goal is to alter the makeup of the soil microbial community so it resembles that of a highly fertile and productive soil.

There are several different ways to accomplish this goal, depending on the situation. Because we cannot see most of the creatures living in the soil and may not take time to observe the ones we can see, it is easy to forget about them. See Table below for estimates of typical amounts of various organisms found in fertile soil. There are many websites that provide in-depth information on soil organisms. Look for a list of these web sites in the Additional Information Resources section of this publication. Many of these web sites have colour photographs of soil organisms and describe their benefits.

Table 4: Weights of soil organisms in the top 7 inches of fertile soil.

Organism	Pounds of liveweight/acre
Bacteria	1000
Actinomycetes	1000
Molds	2000
Algae	100
Protozoa	200

Nematodes	50
Insects	100
Worms	1000
Plant roots	2000

Fertilizer Amendments and Biologically Active Soils

What are the soil mineral conditions that foster biologically active soils? Drawing from the work of Dr. William Albrecht (1888 to 1974), agronomist at the University of Missouri, we learn that balance is the key. Albrecht advocated bringing soil nutrients in balance so that none were in excess or deficient. Albrecht's theory (also called base-saturation theory) is used to guide lime and fertilizer application by measuring and evaluating the ratios of positively charged nutrients (bases) held in the soil. The positively charged bases include calcium, magnesium, potassium, sodium, ammonium nitrogen, and several trace minerals. When optimum ratios of bases exist, the soil is believed to support high biological activity, becomes resistant to leaching, and has optimal physical properties (water intake and aggregation). The plants growing on such a soil are also balanced in mineral levels and are nutritious to humans and animals alike.

Through extensive research, Albrecht determined the desirable percentages of base saturation in the soil. These percentages, he maintained, were optimal for the growth of most crops.

These levels are:

Calcium	60-70%
Magnesium	10-20%
Potassium	2-5%
Sodium	0.5-3%
Other bases	5%

Fertilizer and lime applications should be made at rates that will bring soil mineral percentages into this ideal range. Through this approach, soil pH shifts automatically into a desirable range without creating nutrient imbalances. The base saturation theory also takes into account the effect one nutrient may have on another and avoids undesirable interactions. For example, excess phosphorus is known to tie up zinc.

The Albrecht system of soil evaluation contrasts with the approach used by many state laboratories often called the "sufficiency method."

Sufficiency theory places little to no value on nutrient ratios, and lime recommendations are typically based on pH measurements alone. While in many circumstances base saturation and sufficiency methods will produce identical soil recommendations and similar results, significant differences can occur on a number of soils. For example, suppose we tested a cornfield and found a soil pH of 5.5 and base saturation for magnesium at 20% and calcium at 40%. Base saturation theory would call for liming with a high-calcium lime to raise the % base saturation of calcium; the pH would rise accordingly. Sufficiency theory would not specify high calcium lime and the grower might choose instead, a high magnesium dolomite lime that would raise the pH but worsen the balance of nutrients in the soil. Another way to look at these two theories is that the base saturation theory does not concern itself with pH to any great extent but rather with the proportional amounts of bases. The pH will be correct when the levels of bases are correct.

Albrecht's ideas have found their way onto large numbers of American farms and into the programs of several agricultural consulting companies. Neal Kinsey, a soil fertility consultant of Charleston, MO, is a major proponent of the Albrecht approach. Kinsey was a student under Albrecht and is one of the leading authorities on the base-saturation method. He teaches a short course on the Albrecht system and provides a soil analysis service. His book, *Hands On Agronomy*, is widely recognized as a highly practical guide to its understanding and implementation.

Several firms–many providing backup fertilizer and amendment products–offer a biological-farming program based on the Albrecht theory. Typically these firms offer broad-based soil analysis and recommend balanced fertilizer materials considered friendly to soil organisms. They avoid the use of some common fertilizers and amendments such as dolomite lime, potassium chloride, anhydrous ammonia, and oxide forms of trace elements because they are considered harmful to soil life.

Conventional Fertilizers

Commercial fertilizer can be a valuable resource to farmers in transition to a more sustainable system and can help meet nutrient needs during times of high crop nutrient demand, or when weather conditions result in slow nutrient release from organic resources. Commercial fertilizers have the advantage of supplying plants with immediately available forms of nutrients. They are often less expensive and less bulky to apply than processed natural fertilizers. Not all conventional fertilizers are alike, however.

Many appear harmless to soil livestock but a few are problematic. Anhydrous ammonia contains approximately 82% nitrogen and is applied subsurface as a gas. Anhydrous speeds the decomposition of organic matter in the soil, leaving a soil more compact as a result. The addition of anhydrous contributes acidity to the soil, requiring 148 pounds of lime to neutralize 100 pounds of anhydrous ammonia or 1.8 pounds of lime for every pound of nitrogen contained in the anhydrous. Anhydrous ammonia initially kills many soil microorganisms in the application zone. Bacteria and actinomycetes recover within one to two weeks to levels higher than those prior to treatment. Soil fungi, however, may take seven weeks to recover.

During the recovery time, the bacteria are stimulated to grow and decompose more organic matter due to the high soil nitrogen content. This is why their numbers increase after anhydrous applications. Farmers commonly report that the long-term use of synthetic fertilizers, especially anhydrous ammonia, leads to soil compaction and poor tilth. When bacteria increase and organic matter decreases, aggregation naturally declines because there is no more glue being produced to stick the soil particles together.

Potassium chloride (KCl) (0-0-60 and 0-0-50), also known as muriate of potash, contains approximately 50 or 60% potassium and 47.5% chloride. Muriate of potash is made by refining potassium chloride ore, which is a mixture of potassium and sodium salts and clay from the brines of dying lakes and seas. The potential harmful effects from KCl can be surmised from in the salt concentration of the material. Table below shows that, pound for pound, KCl is surpassed only by table salt on the salt index. Additionally, some plants such as tobacco, potatoes, peaches and some legumes are especially sensitive to chloride. High rates of KCl must be avoided on such crops. Potassium sulfate, potassium nitrate, sul-po-mag, or organic sources of potassium may be considered as alternatives to KCl for fertilization.

Sodium nitrate, also known as Chilean nitrate, or nitrate of soda, is another high salt fertilizer. Because of the relatively low nitrogen content of sodium nitrate, a high amount of sodium is added to the soil when normal applications of nitrogen are made with this material.

The concern is that excessive sodium acts as a dispersant of soil particles, degrading aggregation. The salt index for KCl and sodium nitrate can be seen in Table below.

Table 5: Salt index for various fertilizers.

Material	Salt Index	Salt index per unit of plant food
Sodium Chloride	153	2.9
Potassium chloride	116	1.9
Ammonium nitrate	105	3.0
Sodium nitrate	100	6.1
Urea	75	1.6
Potassium nitrate	74	1.6
Ammonium sulfate	69	3.3
Calcium nitrate	53	4.4
Anhydrous ammonia	47	.06
Sulfate-potash-magnesia	43	2.0
Di-ammonium phosphate	34	1.6
Monammonium phosphate	30	2.5
Gypsum	8	.03
Calcium carbonate	5	.01

Biological and Chemical Aspects of Soil Productivity

Observations such as the accumulative beneficial effects of pasture, and the degradative impacts of successive cereal crops on a coarse-textured sandy loam soil in Australia provide a basis for recommending rotations to maintain sustainability.

There are many beneficial effects from growing legume crops and pastures in rotation. They improve soil structure, biological activity and crop yields, as well as increasing soil organic matter and nitrogen. This Chapter describes those aspects of soil biology, organic matter and soil nutrition central to the observed benefits of crop rotations.

Soil Biology and the Biological Micro-environment

Organic matter, microscopic and macroscopic organisms (e.g., fungal hyphae and invertebrates), detritus from fungi and animals, and bacteria, and biological exudates, all assist in stabilizing soil structure. The role of each part of the biomass differs according to its size. Broadly, large aggregates greater than 250 m m diameter (macro-aggregates), are stabilized by their inherent physical structure, wetting and drying cycles, and organic matter. Micro-aggregates (< 250 m m) are stabilized by live or dead roots, fungi, invertebrates and microorganisms. The populations of soil organisms of all sizes are linked functionally through their roles in the degradation of various forms of organic material. The latter includes live and dead plant material and other live or dead organisms. This shows that animals such as nematodes and some fungi feed directly on live plants while other fungi and bacteria feed predominantly on litter.

Earthworms and other large invertebrates create, and inhabit, burrows and pores, and are very mobile. The most notable of these are termites, which are divided into three groups according to the structure of their nests: those that build mounds (a) above ground, (b) on the soil surface, and (c) below ground. Small arthropods, microfauna and fungi live mostly in larger voids and in association with roots. Foster (1988) reviewed the location of the various types of soil-dwelling organisms and found that fungi, which constitute about 80% of the biomass in many soils, tend to be restricted to the rhizosphere of roots, to larger pores between aggregates and to the surface of aggregates. Bacteria, by contrast, are found on roots in the rhizosphere, in small colonies in the larger micropores, within aggregates and on and within cell debris.

Organic Matter

Both plants and animals provide inputs of organic matter to soils. Once within the soil organic residues can be distinguished on the basis of their chemical structure (e.g., old lignified humic substances that degrade slowly), by their source (plant or animal) or by location.

The standing crop of litter in semi-arid grasslands is usually more than 3 t/ha and in temperate dry steppe may exceed 11 t/ha (e.g., Klemmedson 1989). There has been much debate about the relative contents of organic matter in tropical and temperate soils. Within those wet-and-dry climates that have hot summers assisting rapid decomposition, there is no evidence of inherently lower levels of organic matter in the tropics than in comparable temperate regions (Juo and Payne 1993).

Kowal and Kassam (1978) and Juo and Payne (1993) review the role of organic matter in tropical soils. Here, it is sufficient simply to state that organic matter has various interrelated effects on soil fertility. In particular it should be noted that both chemical and physical effects are of relatively great importance in the soils of the semi-arid tropics because these generally have low cation exchange capacity. The relative importance of litter (crop residue) and manure as inputs of organic matter, varies between cropping systems and spatially within a system. Here, most of the above-ground crop residue is fed to animals, but an equal amount of below-ground crop material enters the soil organic matter pool.

Where alley cropping and agroforestry are practised, values are more variable, but possible inputs could be very significant where the trees, from which the litter is taken, are grown away from the annual crops. If, say, two-thirds of the leaves from leguminous trees are harvested annually, litter values will be substantially higher and the material of better quality than the leaf and stem residue from an annual crop likely to be recycled in the field. Some qualifications, however, should be made. Tree root material is not available for decomposition in the crop field unless it is spatially overlapping (e.g. as an intercrop), in which case the trees will compete with the crop for soil nutrients, water, light and space.

The proportion of animal and human manure used on cropland is more variable. Some farmers have developed stable systems which strongly emphasize the use of animal manure on crops. For example, Norman *et al.* (1982) describe how farmers in northern Nigeria managed to apply 4 t/ha of manure to their heavily-cropped crop land though they had only 3 cows each. Many other farmers do not ensure adequate recycling, either because they are more concerned with livestock management or they do not know the importance of maintaining a 'zero nutrient budget' to replace nutrients removed by the crop. For example, Norman *et al.* (1982) also describe farmers with 10 cows each, who applied only 1.9 t manure/ha to their crops.

Within a cropping system, manuring practice varies with location. There is transference towards the centre of the system. On traditional farms, the area near the household or village is highly fertilized with human and animal manure while more distant fields receive little or no organic matter. Fussell (1992) describes such a traditional 'ring' farming system in semi-arid west Africa. Here, if houses are thatched, the village needs rebuilding or moving every 2 to 4 years. Moving takes advantage of the fertility gradient.

Where the huts are not moved, the fertility gradient becomes steeper with time. Rather than trying to even out fertility by labour-intensive transport of manure, farmers vary the cropping of the fields. Continuous cropping of millet is sustainable close to the hut or village where there is plenty of human and animal manure but crop rotations are essential at the periphery.

Table 6: Biomass production of leaves from multipurpose trees (Source: collated in Young 1989)

Country	Land use	Tree	t/ha/year
Malaysia	Plantation	*Acacia mangium*	3.06
Philippines	Plantation	*Albizia falcataria*	0.18
Costa Rica	Hedgerow intercropping	*Calliandra calothyrsus*	2.76
Philippines	Plantation	*Gmelina arborea*	0.14
Indonesia (Java)	Plantation	*L. leucocephala, A. falcataria, Dalbergia latifolia, Acacia auriculiformis*	3.00-5.00
		Cordia alliodora	2.69
	Plantation crop	*C. alliodora* + cacao,	6.46
Costa Rica	combination	*Erythrina poeppigiana,*	4.27
		E. poeppigiana + cacao	8.18
Nigeria	Hedgerow intercropping	*Cajanus cajan*	4.10
Nigeria	Hedgerow intercropping	*Gliricidia sepium*	2.30
Nigeria	Hedgerow intercropping	*L. leucocephala*	2.47
Nigeria	Hedgerow intercropping	*Tephrosia Candida*	3.07
India	Plantation	*L. leucocephala*	2.30

Closed systems, which rely on maintenance of fertility by recycling organic matter, are not sustainable. Nitrogen levels may be maintained or even enhanced by using legume crops or pastures in rotation but even then further inputs of nutrients are required to replace those removed by the crop.

Fussell (1992) describes the inevitable need to use inorganic nutrients or organic nutrients from off-farm sources such as wastes from cities to complement farm organic matter in the African Sahel: "Although the importance of organic manure in increasing cereal yields is well recognized, it is unlikely that there is sufficient available, on a regional basis, to sustain yields without the use of additional chemical fertilizers."

For example in the Western African savannah 10 t/ha/year of manure are required for sustainable cropping of millet or sorghum...while in northern Nigeria 2.5 t/ha/year are sufficient to maintain yields in most areas."

This four-fold difference largely reflects the integration of livestock with cropping in Nigeria which results in relatively high levels of soil organic matter. Farmers elsewhere, for example in Ghana, give lower priority to crop fertility, which lowers soil organic matter levels and leads to greater nitrogen loss through water erosion when there is little litter cover on the ground.

Irrespective of these differences, however "It is doubtful if either of these levels of application can be achieved, as.. estimates cattle and sheep manure outputs in the northern savannah zone of Nigeria are estimated to be only 1.4 and 0.25 t dry matter per head per annum respectively." Off-farm sources of manure, organic matter generally (e.g., industrial by-products) and soil ameliorants will become increasingly important.

Some, such as sewage from cities, are being researched. Their use in developing countries seems likely to increase, particularly as labour for transporting these materials is cheap relative to the cost of other nutrients such as manufactured inorganic fertilizers. The treatment and efficacy of various off-farm sources of organic matter are outside the scope of this *Bulletin*. It suffices to point out that substantial amounts of material are available.

Table 7: Annual production of organic wastes in the USA and their current use on land

Organic wastes	Total production Million metric tons	Current use on land[1] (%) Percentage of total	
Anumal manures	159	21.8	90
Crop residues	391	53.7	68
Sewage sludge and septage	4	0.5	23
Food processing	3	0.4	(13)
Industrial organic	7	1.0	3
Logging and wood manufacturing	32	4.5	(5)
Municipal refuse	132	18.1	(1)
Total	728	100.0	-

[1]Values in parentheses are rough estimates because of insufficient data.

Table 8: Decomposition constants, k, for tropical legumes. Values were calculated with exponential equation for decomposition using data reported in the literature (Source: Juo and Payne 1993)

Species	Location	Mean annual rainfall	Mean annual temp. °C	k/year
Gliricidia sepium	Ibadan, Nigeria	1250	23-31	8.48
Flemingia congesta	Ibadan, Nigeria	1250	23-31	3.66
Cassia siamea	Ibadan, Nigeria	1250	23-31	2.17
Lonchocarpus cyanescems	Ibadan, Nigeria	1250	23-31	8.87
Inga vera	El Verde, Puerto Rico	4000	22	1.65
Inga sp. And *Erythrina* (mixed)	Caracas, Venezuela	1200	20	3.01
Erythrina sp. (mixed with non-legumes)	Caracas, Venezuela	1200	20	3.81
Inga edulis	Yurimaguas, Peru	2200	26	0.91
Cajanus cajan	Yurimaguas, Peru	2200	26	1.45
Erythrina sp.	Yurimaguas, Peru	2200	26	3.72

The rate of breakdown of litter and other organic materials both determines and depends on the populations and activities of organisms in the soil. It also determines the extent to which minerals taken up by a crop are released from its organic residues and made available as inorganic ions to the subsequent crop.

There are numerous mathematical models of organic matter degradation. Almost all assume that the rate of degradation decreases with time, as the more soluble, digestible and accessible material is selectively and progressively degraded by soil organisms. This selectivity is often arbitrarily described by recognizing several 'pools' of organic matter and, with time, 'passing' material from labile to less-labile compartments. In practice its physical and chemical composition changes gradually with time, as does its physical binding with soil particles and its relation to the predators that feed on it.

The rate of decomposition of leaf litter depends on its environment, particularly on soil temperature and soil water. Both these affect the physical breakdown of the litter and determine the population and activity of soil animals and fungi that feed on it. Decomposition also varies with plant type and age of litter, being slower for heavily lignified material. The specific properties of litter from different species, and the generally exponential form of litter decay (the rate of decomposition slowing with time) lead to values that suggest half-lives of litter ranging from 1 to about 10 years. Because of the importance of temperature in determining decomposition, rates decline from tropical to temperate locations within wet-and-dry climates.

Plant species, water and temperature are not the only factors affecting the speed of litter decomposition and mineral cycling. Management plays a role. There are four ways for managing crop residues: (a) stubble mulch in which residues are left standing; (b) surface mulch, where above-ground residues are cut and left on the top of the soil after harvest; (c) incorporation by ploughing; and (d) cut-and-carry, in which surface residue is removed and (if not used for livestock or thatching, etc.) returned as a surface mulch about planting time for the subsequent crop; this is usually combined with ploughing of below-ground residues.

Two of the above treatments involve the fragmenting of root residues. All of them affect a range of soil properties. The degree of contact between the residue and the soil (and its organisms) and other unexplained factors affect the rate of decomposition.

Suggests that the rates of release of minerals from organic matter, particularly litter, are a significant element in nutrient cycling and crop productivity in semi-arid environments. Such climates have a rainy season in which rainfall exceeds evaporation, so there is a risk of mineral loss through runoff and mineral loss and acidification by leaching. During the dry season there is potential loss of nitrogen as ammonia. This suggests the need to manage litter so most decomposition takes place when the subsequent crop is growing rapidly and is able to take up the nutrients released. It is wise at the same time to maintain ground cover if possible. Manipulation of litter requires labour, which may already be in maximum demand prior to sowing.

Soil Organisms Associated with the Crop

Bacteria and Nitrogen

Cropping in dryland regions needs nitrogen to be economically successful (e.g., Keating *et al.* 1991). Two sources of nitrogen are from organic matter and from nitrogen-fixing bacteria associated with plant roots. *Bradyrhizobium* and *Rhizobium* species infect plant roots forming galls or nodules, and fix nitrogen from the soil atmosphere directly to the plants. Locally-adapted, heat-tolerant strains survive from crop to crop in wet-and-dry climates and, whether established by natural colonization or by inoculation of the crop seed at sowing, they subsequently fix variable quantities of nitrogen. Where bacterial infection is effective, the bacteria commonly fix between 70-100% of the nitrogen used by the crop, though the proportion is lower when the crop is given inorganic fertilizers.

Table 9: Effect of soil mineral N and N fertilizers on crop N productivity and the proportion (P) and amount of crop N derived from N_2 fixation (Source: Peoples and Craswell 1992)

Species	Location	Level		Total crop	N_2 fixed	
		Soil mineral N (kg N/ha)	Fertilizer N (kg N/ha)	N (kg N/ha/crop)	Proportion	Amount (kg N/ha/crop)
Groundnut	India	-	0	196	0.61	120
			100	210	0.47	99
			200	243	0.42	102
Chickpea	Australia	10 (to 120 cm)		114	0.85	97
		326		184	0.17	33
			0	109	0.80	87
			50	110	0.55	60
			100	104	0.29	30
Soybean	Australia	70 (to 120 cm)		230	0.34	78
		260		265	0.06	16
	India	-	0*	63	0.29	18
			100	148	0.26	28
		-	0**	89	0.48	43
			100	115	0.24	28
Common bean	Kenya	-	10	149	0.39	58
			100	158	0.10	16
Cowpea	Kenya	-	20	116	0.53	62
			100	137	0.08	11
	India	-	0	163	0.77	125
			100	138	0.67	92
			200	172	0.33	57

* Uninoculated.

** Inoculated.

The extent of the effectiveness of infection of legume crops in the wet-and-dry tropics needs to be surveyed. Temperate research indicates that nitrogen fixed by bacteria ranges from 20 to 120 kg N/ha in a growing season for annual crops. In the semi-arid tropics, amounts of nitrogen fixed per hectare range from none, where nodulation is ineffective, to 16 kg N in soybean naturally colonized by rhizobia, to 84 kg N when inoculated. Nitrogen fixation in soybean and groundnut of 50-70 kg N/ha/growing season is reported in Senegal (Gigou *et al.* 1985). Nitrogen-fixing bacteria associated with tree legumes can fix similar quantities.

Rhizobia rarely fix all the nitrogen used by the crop and a substantial amount of nitrogen is removed from the field as grain, so the net benefit from symbiosis varies.

For example, if rhizobia fix 50% of the crop's requirement and 60% of above-ground nitrogen is in the grain at harvest, soil nitrogen is depleted irrespective of the amount of nitrogen fixed or the grain yield. Peoples and Craswell (1992), studying 21 cases of tropical grain legumes, found positive nitrogen balances in 15 and negative ones in five. Similarly, many experiments with grain legumes in southern, semi-arid Australia show that rhizobial associations give net benefits in more than three-quarters of cases and that the benefit from lupins is substantially greater than that from field peas, largely because a smaller proportion of nitrogen is harvested and removed in lupins.

Fungi, Algae and Nutrients

Various fungi facilitate uptake of nutrients by plants, particularly phosphorus. Numerous fungi live in close association with plant roots. One group, vesicular arbuscular micorrhizal fungi (VAM), form both vesicles and arbuscules (knot-like structure) on the surface and within the root. They also colonize soil animals including earthworms and woodlice.

They are most prolific in the topsoil, to about 10 cm depth (Habte 1989). They facilitate uptake of nutrients, particularly phosphorus from soils low in that element.

They provide some protection to the host plant, their presence being associated with decreased colonization by pathogens. Ellis *et al.* (1985) also found that wheat plants inoculated with VAM were more drought-tolerant than plants without VAM. Importantly, comparisons of conventional cropping systems using inorganic fertilizers and herbicides with organic systems not using herbicides have found much higher levels of infection of crop roots by these beneficial fungi in the organic system (Ryan *et al.* 1994). This suggests that management for desired, associative soil organisms, should become a part of future sustainable cropping.

Table 10: Nitrogen fixation by trees and shrubs. Values are per growing season or per year unless the number of months is given in brackets (Various sources, compiled by Young, 1989 and Peoples and Craswell 1992)

Species	N fixation (kg N/ha/year)
Acacia albida	20
Acacia mearnsii	200
Allocasuarina littoralis	220(?)
Casuarina equisetifolia	60-110
Coffee + *Inga* spp.	35
Coriaria arborea	190
Erythrina poeppigiana	60
Gliricidia septum	13
Inga jinicuil	35-40
Inga jinicuil	50
Inga jinicuil	35
Leucaena leucocephala	100-500
Leucaena leucocephala (in hedgerow intercropping)	75-120
Leucaena leucocephala	100-13 (6)
Prosopis glandulosa	25-30
Prosopis glandulosa	40-50
Prosopis tamarugo	200
Rain forest fallow	400-100
Mature rain forest	16
Alachynomene afrospera	130 (2)
Glyricidia sepium	100 (3)
Sesbania spp.	125-140
	119-188 (2)
	140-290 (2)
Xalliandra colothyssus	11 (3)

Other organisms associated with or infecting plants include endophytic fungi which affect plant growth rates and confer resistance to some insects. Cyanobacteria, which belong to the blue-green algae, are believed widespread in the semi-arid tropics and contribute 0.5 to 15 kg N/ha/year. *Azospirillum, Klebsiella, Enterobacter* and other species are free-living but largely inhabit the root rhizosphere; their distribution and role is still to be defined for cropping systems in wet-and-dry climates.

Actinorrhizal associations are known between *Casuarina* and *Frankia* sp., but these are not exploited. It is claimed that such associations can fix as much as 60 kg N/ha/year. There are no doubt other symbioses yet to be discovered and exploited.

Weeds, Crop Pests and Diseases

Weeds, pests and diseases all compete with or directly reduce the vigour of crops. Many pests and diseases are soil-borne. Weed life-cycles depend on replenishment of the soil seed bank and survival of the seeds against natural decay, predation by soil animals and depletion by human management, particularly cultivations. Ecological weed control thus aims to minimize recruitment of new seed into the soil as a long-term strategy as well as trying to reduce artificially the size of the weed seed bank in the soil.

Table 11: Frequency (%) of VAM fungi in soil-dwelling macro-invertebrates sampled from natural and agricultural ecosystems in Ohio (data are combined from 1986 and 1987 samplings) (Source: Rabatin and Stinner 1989)

Taxa	Ecosystem			
	Conventional tillage maize	No-tillage maize	Pasture	Old field
Lumbricidae (earthworms)	25.0	83.3	50.0	75.0
Isopoda (woodlice)	100.0	35.7	64.7	36.8
Carabidae (ground beetles)	2.1	19.8	14.5	12.8

Weed life cycles and strategic weed control are well understood for temperate wet-and-dry climates. The weed populations were measured under the contrasting conditions of slash-and-burn and terrace cultivation. Seed recruitment to the soil seed bank was higher under terrace cropping than under slash and burn but fewer survived.

It is generally accepted that there are links between weed populations, their competitiveness, and soil fertility. A good example of this is the various species of witch weed (*Striga hermonthica, S. generioides, S. asiatica* and approximately 25 other species). These species infect the roots of millet, sorghum, maize and sugar cane. *S. generioides* also infects legumes and tuberous crops.

They are a serious nuisance and are widespread in semi-arid Africa. It is estimated that they currently infect 21 million ha and a further 44 million ha are at risk. Their spread and increased aggressiveness is considered to be due largely to declining fertility and soil carbon contents.

The incidence of insect pests and crop diseases depends on the complexity of the cropping system and the types of crops grown. It depends too on management factors, such as the timing and type of tillage and the treatment of the crop residues. These aspects are so complex they warrant a Bulletin to themselves, so they are only touched upon here.

The incidence of pests and diseases is generally believed to be less frequent and/or intense in diverse cropping systems. Data on insects are relatively scarce but Andow (1983) is frequently quoted as showing a positive relationship between insect pest populations and the practice of monoculture. His experience, however, is mainly with cotton, a unique crop, and wheat, though he gives a few examples from maize and rice.

Table 12: Viruses for which disease incidence is lower in more diverse cropping systems (Source: various, collated in Power 1990)

Pathogen	Type[1]	Vector
Alfalfa mosaic virus	N	Aphid
Bean common mosaic virus	N	Aphid
Bean yellow mosaic virus	N	Aphid
Beet yellows virus	P	Aphid
Cauliflower mosaic virus	N	Aphid
Chlorotic mottle virus	N	Aphid
Maize stunt spiroplasma	P	Leafhopper
Cucumber mosaic virus	N	Aphid
Groundnut rosette virus	P	Aphid
Pepper veinbanding mosaic virus	N	Aphid
Tomato yellow leaf curl virus	P	Whitefly
Turnip mosaic virus	N	Aphid

[1] Type of pathogen transmission system is given as P (persistent) or N (non-persistent).

Power (1990) shows that the incidence of virus diseases is less in complex cropping systems. He suggests that disease transmitted by insects is less common in multiple-cropping than in monocultures because aphids and some other insects settle on plants in response to visual cues that do not enable them to distinguish between host and non-host plants. In this way polyphagous insects can acquire a pathogen from an infected plant and transfer it when feeding on a non-host plant. If the insect is preferentially attracted to the non-host crop, it may be speculated that this crop acts as a 'trap' or 'protection' crop.

Power (1990) reviews situations where intercropping or the inclusion of a second, barrier species has led to lower disease levels. In contrast, bacteria and fungi in the root region and rhizosphere are reported to be slightly fewer in a coconut monocrop than in a multi-storied mixed cropping system (Bopaiah and Shetty 1991). Differences in populations, and in their varied (positive or negative) effects on the food crop, are no doubt best understood through analysis of the feeding habits and predation of the organisms themselves, rather than by making superficial generalizations about their numbers. Fick and Power (1992) and Hearn and Fitt (1992) review pest management generally and pests and diseases in cotton crops specifically.

Inorganic Nutrition

Inorganic nutrients occur in soil as ions and minerals, e.g., as oxides, silicates and phosphates, both adsorbed onto the surface of clay particles and organic matter, and in solution.

A large proportion of some nutrients, most notably nitrogen, is found in organic matter in all but the most highly degraded soils, so organic matter and the organisms associated with it are given prominence in this chapter.

Clay particles, because of their crystalline structures, carry an inherent electrical charge. This results in attractive forces (mainly van der Waals forces) and repulsive forces (electrostatic forces) which give clay species their particular characteristics. The inherent surface charge also causes a layer of associated ions to align next to the solid particles forming a so-called diffuse double layer because it consists of a relatively inexchangeable layer (the Stem layer) closest to the surface of the particle and an outer, readily exchangeable layer, of varying thickness, called the diffuse layer. Sposito (1984) explains this more fully.

In the present context, it is appropriate to describe briefly four aspects of the physico-electrical structure of soils. These are: the inherent exchange capacity; the linkage between soil particles and soil organic matter, and its effects on soil stability; ions in solution and their solubility; and the ion imbalances (from the perspective of crop growth) associated with acidity and salinity.

Table 13: Representative cation exchange capacities (in mol_c/kg) of surface soils (Source: Sposito 1984)

Soil order	CEC
Alfisols	$0.1\ 2 \pm 0.08$
Aridisols	0.16 ± 0.05
Entisols	0.13 ± 0.06
Histosols	1.4 ± 0.3
Inceptisols	0.19 ± 0.17
Mollisols	0.22 ± 0.10
Oxisols	0.05 ± 0.03
Spodosols	$0.11\ \pm0.05$
Ultisols	$0.06\ \pm0.06$
Vertisols	0.37 ± 0.08

Adsorption is the net accumulation of matter at the interface between a solid phase and an aqueous solution. Readily exchangeable ions are those so loosely adsorbed to clay or other particles that they are replaced easily by leaching with an electrolyte solution.

The ion exchange capacity of soil is the number of moles of adsorbed ion charge which can be displaced from a unit mass of soil; in most applications, this refers to readily exchangeable ions. Sposito (1984) notes that 'Much controversy exists over the surface chemical significance of ion exchange capacities'. The maximum surface charge measurement indicates a soil's potential to adsorb ions while its actual capacity, which is more relevant agriculturally, has a lesser value. It is notable that though the CEC of each order ranges widely, the predominant soils in wet-and-dry climates have low CECs.

The readily exchangeable proportion of the CEC varies with soil pH: the proportion available to readily-exchangeable bases decreases from about 1 at pH 8 to 0.5 at pH 6 to perhaps 0.2 at pH 4.5. Below pH 6 an increasing proportion is taken up by various aluminium ions and complexes, which are variably toxic to plants.

The electrical conductivity and CEC of a soil are related to its clay content (e.g., Rhoades 1990a). Similarly, because of the electrical charge of the clay particles, a high but variable percentage of the soil organic matter is bound to them. That it may be as high as 90% and illustrates the two postulated main ways that organic matter is bonded to clay. These are weak anion exchange and strongly-held ligand exchange which is a form of chemical bonding.

It follows from the electrically-charged nature of clay and organic matter and the high likelihood of their association that both contribute to the nutrient-holding capacity and stability of soils. This intimacy led Pieri (1992) to propose that structural stability, at least of the African soils he studied, could be described by a critical value of S greater than 9, where S is the ratio of organic matter to clay plus silt, expressed as a percentage.

Table 14: Proportion of soil organic carbon in the clay-organic complex

	Percent organic carbon in soil	Percent soil carbon in clay-organic complex[1]
Podzol	1.5	18
Solonized brown soil	1.5	77
Grey clay soil	1.1	91
Terra rossa	2.8	82
Groundwater rendzina	5.4	69
Black earth	1.8	82
Krasnozem	4.9	90
Red brown earth	2.5	66

[1] Defined as the material sinking when the soil was ultrasonically dispersed in an organic liquid of density 2 g/cm^3.

Plants take up most minerals as inorganic nutrients from the soil solution (except legumes, which may fix dinitrogen gas directly from the soil air). An element is available (to the plant) if it is present as, or can be transformed into, a free ion, and it is within the plant root zone. Most elements move within physical proximity of roots through soil water movement and into the plant through evapotranspiration.

Diffusion along concentration gradients is important for less-mobile ions such as phosphorus, particularly where soil solution concentrations are weak and root densities are high (i.e., the transport path is short). Sposito (1984) and others give calculated values for the diffusivity of nutrients. Diffusion times range from 1 day for an ion to move 3 mm (which is comparable with the time it would take to move by convection in the mass flow of water) for nitrate to about 200 days for potassium, magnesium and molybdenum, and to thousands of days for other nutrients.

The concentrations of nutrients in solution fluctuate daily and seasonally. The most dynamic are nitrate and ammonium ions, which are interconverted by bacteria. Fluctuations can be explained by the effects of temperature and soil water on mineralization (breaking down organic matter to release ammonium), immobilization (the reverse) and nitrification (conversion of organic matter to nitrate, which is stable and highly soluble), and by the effects of rainfall leaching nitrate to depth.

In seasonally wet-and-dry cropping systems, there is a flush of nitrate following the start of the wet season and nitrate-N may accumulate in the topsoil by capillary rise of water during the dry season. Within the major seasonal patterns driven by soil water, Rochester *et al.* (1991) found that the soil nitrate cycle lagged three months behind changes in temperature.

Table 15: Relative salt tolerance of various crops at emergence and during growth to maturity (Source: Rhoades 1990b)

		Crop Electrical conductivity of saturated-soil extract	
Common name	**Botanical name**	**50% yield (dS/m)**	**50% emergence (dS/m)**
Barley	*Hordeum vulgare*	18	16-24
Cotton	*Gossypium hirsutum*	17	15
Sugarbeet	*Beta vulgaris*	15	6-12
Sorghum	*Sorghum bicolor*	15	13
Safflower	*Carthamus tinctorius*	14	12
Wheat	*Triticum aestivum*	13	14-16
Beet, red	*Beta vulgaris*	9.6	13.8
Cowpea	*Vigna unguiculata*	9.1	16
Alfalfa	*Medicago sativa*	8.9	8-13
Tomato	*Lycopersicon lycopersicum*	7.6	7.6
Cabbage	*Brassica oleracea capitata*	7.0	13
Corn (maize)	*Zea mays*	5.9	21-24
Lettuce	*Lactuca sativa*	5.2	11
Onion	*Allium cepa*	4.3	5.6-7.5
Rice	*Oryza sativa*	3.6	18
Bean	*Phaseolus vulgaris*	3.6	8.0

Low soil pH affects plant roots directly because of the effects of the hydrogen ion concentration on root membrane integrity and exchange capacity. Acidity also affects roots indirectly in two ways. It alters the availability of ions in the soil solution (possibly making available toxic aluminium species, relatively toxic to plants). It also affects mineralization through the protons competing with cations for dissolved ligands and surface charged groups.

Soil pH also affects microorganisms, and thus the speed of transformations, for example, those between nitrate and ammonium. Poor plant growth on acid soils may thus be caused directly by hydrogen ions, by toxicities of aluminium or manganese, or through deficiencies of calcium, magnesium, potassium, phosphorus, nitrogen or trace elements. Chase *et al.* (1989) describe these relationships for sandy Sahelian soils. Variation in pH across distances of 15 m can be as much as pH 4.5 to 7.5, with associated decreases in aluminium and hydrogen ions, and increases in crop productivity. The critical pH at which crop growth is affected varies with crop, cultivar and soil type. Critical levels may be as high as pH 5 to 5.5 for less tolerant plants in soils with soluble sources of aluminium but otherwise can be as low as pH 3.9 to 4.

Saline soils and saline-sodic soils have electrical conductivities greater than 4 dS/m. The saturated soil extract of saline soils contains more than 15% exchangeable sodium; that of sodic soil less than 15%. The pH values of both kinds of soils range between 7 and 10.5 directly reflecting the amount of sodium bicarbonate present, this being completely dissolved in moist soil.

Field Indicators of Biological and Nutritional Problems

In this chapter, whilst reviewing relevant recent work on soil organic matter, soil biology and inorganic soil nutrients, a non-traditional perspective has been taken by placing emphasis on soil organic matter and the close chemical and functional relationship between organic matter and soil particles. It is therefore consistent to suggest that the most obvious indicator of biological and nutritional problems is the absence of, or reduction in, organic matter. Rates of loss of organic matter independent of erosion tend to be slow, but important as they are cumulative.

Larson and Pierce (1991) propose that minimum, analytically-gathered data sets are essential for monitoring soil sustainability. They include two measures of organic matter among the ten attributes that they consider essential. Once the minimum data set is constructed

(Larson and Pierce 1991), each soil attribute is determined for a reference time (T0) and the change in soil conditions can then be measured over time (T1), which they suggest should be one to ten years.

While suggesting that agreed repeated observations or measurements are essential, it is preferable to empower farmers to take responsibility for recording their own observations of surrogate measurements rather than giving responsibility for analytically-based measurements to outsiders. Policy-makers need to address the need for analytical or surrogate assessments of soil sustainability. A first step would be to ask farmers for their observations of soil characteristics and their attitudes and concerns about soil productivity.

In the case of soil organic matter measurements, it is believed that the limited availability of expertise, the costs and the inherent sampling errors make direct measurement of organic carbon neither cost-effective nor informative. It is more likely that surrogate measures are adequate and, if sufficiently simple and cheap, have some likelihood of being used. Pieri (1992) suggests that a bleached (possibly brittle) soil surface and plant deficiency symptoms are useful surrogates for loss of organic matter, low CEC and soil nutrient imbalances.

Indicators of plant nutrient deficiencies include: colour (e.g. redness for potassium deficiency); paleness (a more general symptom, but especially obvious in nitrogen deficiency); and small leaves and plants. Other symptoms, such as leaf curling and accelerated dropping of older leaves, may also be helpful but might equally indicate water deficits. There are several publications with photographs of nutrient deficiencies; these, however, might require further tailoring to specific combinations of crops and soils.

Soil acidity is most obviously indicated by stunted plants, by bare ground, and by better relative growth of acid-tolerant weeds in a sensitive crop. Salinity too is indicated by vegetation changes such as trees dying for no apparent reason or an increase in the proportion of salt-tolerant herbs. Other symptoms of salinity include: waterlogged or bare soil; livestock congregating and licking the soil surface for salt; visible salt crystals; the smell of salt; and clear catchment water because salt settles sediment.

Management for Maintenance of Soil Biology and Nutrition

The aim of management should be to create balanced organic matter and mineral budgets. It should ensure that, over several years (a complete crop rotation), soil organic matter is not depleted and that nutrients added equal or exceed those removed by cropping or lost in various ways.

When managing organic matter farmers should recognize that the effects of animal and human manure, sewage sludge and plant residues last longer than those of green manure crops. Green manures, though valuable, usually last only one or two seasons because they are incorporated before they are mature and lignified.

Juo and Payne (1993) advise: 'In spite of the many proven benefits of soil organic matter, its management and recycling in an intensified, modem agro-ecosystem must necessarily revolve around two fundamental characteristics of the system, namely, the availability of organic material at the farm level, and the economic incentive for conserving and recycling organic matter'. To this may be added consideration of the benefits and costs of treating and transporting human wastes from cities. On-site organic matter, organic material brought in from outside and industrially-produced fertilizers will be used according to their benefit cost ratios and the attitudes of governments and crop managers.

Tillage reduces the frequency of VAM, at least in invertebrates. Tillage reduces earthworm populations 10-30-fold, both killing them directly and destroying their burrows (Brust et al. 1986). Haines and Uren (1990) show that differences in earthworm populations between tillage treatments and stubble burning are reflected by 50% differences in numbers of soil pores.

Tillage also effects the location, numbers and activity of microorganisms within the soil, depending on the type of tillage. For example a mouldboard plough that inverts topsoil has more effect than minimal tillage. Doran (1987) found microbial biomass and potentially-mineralizable nitrogen to be 54% and 37% higher respectively in the top 7.5 cm of no-till than ploughed soil.

These would be critical differences if translated to the semi-arid tropics (Keating et al. 1991). Microbial and fungal biomasses may be 20-70% higher after 1 and 2 years no-till (Doran 1980). However, conventional tillage, by burying topsoil, may cause microbial populations and their activities to be higher than under no-till at depths of say 7.5 to 15 cm (Doran 1980). Differences in soil organisms (e.g., microbial biomass) under differing tillage treatments are seasonal, generally being greatest in the period when the weather is most favourable for soil organism activity and plant growth (Carter and Mele 1992). Rasmussen and Collins (1991) summarize many data on the impact of tillage on soil properties and conclude that stubble-mulching with zero tillage conserves up to 2% more organic matter per year than ploughing.

Table 16: Micro-arthropod populations in top 10 cm of clay soil 1 and 2 years after adding organic fertilizers in Italy. Additions were made to achieve 4% organic matter and inorganic N and P were to make all treatments equivalent to the manure (Source: Fratello *et al.* 1989)

Taxon	Control	Fowl manure	Sludge	Green manure	Straw
One year after application					
Collembola	13 209	31 635	13 653	17 205	24642
Entomobryidae	2 109	4 329	2 109	222	2 331
Onychiuridae	1 221	333	2 109	222	1 221
Isotomidae	9 990	26 529	9 324	16 650	20 313
C. coecus	8 214	24753	7 659	10 101	18 204
C. thermophilus	0	333	888	0	0
I. Minor	111	111	0	4 995	0
F. Parvulus	0	111	0	0	0
Sminthuridae	0	222	0	111	222
Neelidae	0	111	111	0	555
Acarina	65 046	42 513	35 853	43 068	70263
Pauropoda	3 108	1 221	3 108	2 331	7 104
Microarthropods	83 361	80475	54834	65 157	103 674
Two years after application					
Collembola	4773	7 104	8 769	5 328	5 550
Entomobryidae	0	999	444	222	111
Onychiuridae	444	444	*111*	222	0
Isotomidae	2 886	1 998	6 771	4 107	4 551
C. coecus	1 887	999	2 442	3 108	3 885
C. thermophilus	0	222	3 996	222	111
I. Minor	222	222	0	333	0
F. Parvulus	333	222	0	0	333
Sminthuridae	1 221	3 552	1 221	777	888
Neelidae	222	0	0	0	0
Acarina	12432	28083	24753	27 639	13 764
Pauropoda	999	111	1 554	999	777
Microarthropods	19425	35 964	36 963	35 187	21 090

Cropping patterns and rotations affect soil nutrition. Diversity of cropping increases the number and variety of soil organisms and reduces pests and diseases.

Soil acidification and salination are extreme cases of nutrient imbalance and, unlike other deficiencies, cannot be corrected simply by adding mineral fertilizers. The management techniques to reduce acidification are: (a) to reduce the net proton production of the system by minimizing nitrate leaching (a special problem in seasonally wet-and-dry climates); (b) to avoid using ammonium fertilizers and reduce the accumulation of organic matter; (c) to lime the soil. Liming is a high cost option while adoption of perennial species (to avoid seasonal accumulation of organic matter and nitrate leaching) is less costly. By contrast, salinity is usually managed by strategic leaching. This is done with the sparing use of low-salt water when both soil moisture and the water table are low.

Soil Bulk Density

Soil Bulk Density Determination

Soil weight is referred to as soil bulk density. Density is the mass of material contained within a given volume. This is the idea that a given size of box may be heavy or light, depending upon what kind of material it contains. If the box were filled with wood, it would be light when compared to having it filled with lead. The weight of water is the reference for density measurements: 1 gram of water=1 cubic centimeter (cc), & 1 cc water=1 ml. & 1 cubic foot of water=62.4 lbs.

The bulk density of a soil is the mass of dry soil per unit of "bulk" (total volume of soil or soil particles & pore space).

B.D. = mass of oven dry soil (grams) ÷ total volume of soil (cm^3)=grams/ cc

The bulk density takes into account the total soil volume (the space occupied by the solid particles plus the space occupied by the air of the pores or pore space). To determine the mass of the soil—since air does not have any significant weight—we can just weigh the oven dry soil on a balance. The volume of the soil can be determined by pouring the soil into a graduated cylinder and measuring the volume that it occupies.

The problem with this procedure is that structural aggregates may be crushed or compacted once they are removed from the soil and placed into the cylinder. A better procedure is one in which the aggregates could be removed from the soil and frozen exactly the way they were in the soil. Plastic fixatives enable us to do this. A large aggregate has been removed from the soil and is being fixed by dipping it into a saran solution. This will make the clod impervious to water. The weight of our clump of soil can be obtained by hanging it on the balance or placing it on the metal weighing tray.

The volume of the clod also needs to be determined. This can be obtained by weighing the sample again, only this time the sample will be in water. The weight of the sample now will be less, since the sample will be buoyed up by the amount of water it is displacing.

Archimedes most famous theorem gives the weight of a body immersed in a liquid, called Archimedes' principal. If you want to know more about the mathematician who determined that this procedure would work, go to Archimedes.

Thus, by subtracting the weight of the clod in water from its weight in air, we obtain the weight of the water displaced by the clod, which equals the volume of water displaced by the clod (because 1 cm^3 of water=1 gm. of water) and is equal to the volume of the soil clod.

Clod Method of BD Determination =Mass of Clod ÷ Volume of water displaced

Volume of water displaced=weight of water or the (gms clod in air)-(gms clod in water)=(grams or volume of water displaced)

Cylinder method of BD Determination=Mass of oven dry soil (gms) ÷ total volume of soil (cm^3)

Another method of determining bulk density is by using soil cores. These are obtained with a probe that forces a cylinder into the soil. The soil can be removed from this cylinder and cut into sections and placed in another cylinder for transportation back to the laboratory. Care needs to be taken to avoid compacting the soil during and after obtaining the core.

BD of soil (using core method)={(mass of soil + mass of core)-(mass of core} ÷ (volume of core)

volume of core=Pi (3.1416) r^2 × h

r=radius of core; h=the height of core.

For the cores in the lab, r=2.45cm (½ of Diameter); h=5 cm

V=3.1416 × 2.45^2 × 5=94.3 cm^3;

Bulk density is a fairly easy determination that will yield information about the soil that will be significant for determining the soil's potential for plant growth or a building foundation. Remember it is the oven dry mass/volume soil. or g/cc.

Significance of Bulk Density

The bulk density of the soil will play an important role in determining if the soil has the physical characteristics necessary for plant growth, building foundations or other uses.

From the laboratory investigations you will obtain very high bulk densities for the Bt and sandy soil. The weight of the soil is important if you are going to be lifting it or hauling it long distances. One of the main reason for sod farms to be on organic soils is to reduce the cost of transportation because the organic soil is so much lighter than mineral soils. The organic sod is easier to handle. Most sod farms in Minnesota are on the peat soils north of the Twin Cities.

Sometimes we are interested in the weight of an acre of soil for erosion comparison purposes. Erosion of 5 tons per acre sounds like a lot. However the weight of an acre furrow slice on average is 2,000,000 lbs., so 5 tons per acre seems like a small amount. Five tons per acre is only about 34 thousandths of an inch thick (.034 inches). However, many areas have lost 10 to 100 times that amount. (Note: an acre furrow slice, or AFS, is a three-dimensional volume of soil which is one acre in area and 7 inches deep.)

Weight per AFS=BD   Volume

Example: Volume of soil=Soil depth × 43560 ft.^{2} (area of 1 acre)

For conversion of g/cc to lbs/ft^3, multiply BD × 62.4 lbs/ft^3 (weight of ft^3 water)

How many tons are lost if a soil with a B.D. of 1.2 g./ cm^3 erodes 5 inches per acre?

answer: (1.2 × 62.4)=(weight in lbs/ft^3) X Volume: (where Volume=(5 in. ÷12 in/ft) × 43560 ft)^2 or (74.88 lbs/ft^3) × 18,150 ft^3=1.3 million lbs or divide by 2000 lbs / ton or weight=679 tons.

Soil Factors Impacting Tillage

Another factor similar to carrying soil is its potential to be moved short distances, such as in plowingor rototilling. The soil that is heavier will be more difficult to move. However, another factor is the ability of the soil to stick together, or the soil's consistence.

Clay soil is noted for its stickiness and large energy requirements in tillage. Farmers refer to it as "heavy," but they really mean it is difficult to plow, not that it has a high bulk density; clay soils generally will have a lower bulk density than sandy soils (clay=1.3 g/cc vs sand=1.6g/cc). Sandy soils have a higher bulk density, but are easier to plow since they have weaker consistence. Thus they are often referred to as "light soils." The lower B.D. of clays is due to their better aggregation. Rototilling the soil reduces the bulk density by "fluffing" the soil. For a look at tillage implements go to Tillage Implements

Soil Consistence

Soil consistence is the soil's ability to cohere or stick together. The soil's consistence may be evaluated at three moisture conditions: air dry, moist, and wet. Moist consistence is evaluated by placing the soil between the thumb and forefinger and gently applying pressure. The ease with which a ped can be crushed determines the consistency.

Terms commonly used to describe moist consistence are:

Loose-Non-coherent when dry or moist; does not hold together in a mass.

Friable-When moist, crushes easily under gentle pressure between thumb and forefinger and can be pressed together into a lump.

Firm-When moist, crushes under moderate pressure between thumb and forefinger, but resistance is distinctly noticeable.

Plastic-When wet, readily deformed by moderate pressure but can be pressed into a lump; will form a "wire" when rolled between thumb and forefinger.

Sticky-When wet, adheres to other material and tends to stretch somewhat and pull apart rather than to pull free from other material.

Hard-When dry, moderately resistant to pressure; can be broken with difficulty between thumb and forefinger.

Soft-When dry, breaks into powder or individual grains under very slight pressure.

Cemented-Hard; little affected by moistening.

In air dry conditions, the resistance to rupturing when rubbed is measured. At intermediate moisture content, the soil's resistance to shearing forces by thumb and finger is noted. In the wet condition, its plasticity—ability to be molded and stickiness—are measured.

One significance of the B.D. of a soil is the amount of surface area a soil has. The more surface area, the more ability to retain water and nutrients. Notice that the two soils pictured below weigh the same, but are significantly different in surface area. Which soil has the higher B.D.?

Compaction, Porosity and Soil Temperature

Soil Compaction

Soil compaction works something like the soil particles seen below. When the soil is randomly fluffed, as on the left, the amount of pore space will be 45 to 55%. When we push the soil particles closer together, we remove the pore space, and the bulk density will increase. Compaction increases the bulk density by reducing the pore space.

Compaction can also result in the changing of the proportion of pore sizes. Notice in the compaction diagram that compacted soils not only have a lower total pore space (resulting in higher bulk density), but also less macro pores and more micro pores. This will often result in excess water retention.

When we walk on the soil we compact it. Eventually the compaction prevents the growth of plants. This trail TRAIL behind the Soils Building shows the effect of compaction and samples from the soil will be used in the lab to determine the amount of compaction.

Compaction changes the amount and size of pores.

Deep plowing can reduce compaction below the plow zone.

C.O.L.E.

One of the important engineering properties of soils is related to the clay content and consistency and is called the coefficient of linear extensibility or C.O.L.E. It is a measure of the shrink-swell potential, or the volume change of the soil with changes in moisture content. If the C.O.L.E. exceeds.09, significant shrink-swell potential can be expected.

Soils with high C.O.L.E. values can cave in basements, displace walls, and break gas or water pipes. Knowing which soils have high C.O.L.E. values is an important step in selecting sites for construction activities. Compacting soils reduces the ability of soils to take on water and reduces their shrink/swell potential.

C.O.L.E.=(Length Moist ÷ Length Dry)-1

Soil Compaction and Building

Soil compaction is defined as the method of mechanically increasing the density of soil. In construction, this is a significant part of the building process. If performed improperly, settlement of the soil could occur and result in unnecessary maintenance costs or structure failure. Almost all types of building sites and construction projects utilize mechanical compaction techniques

Soil density is used almost exclusively by the transportation industry to specify, estimate, measure, and control soil compaction. Soil density can be easily determined via weight and volume measurements. The objective of compaction is to stabilize soils and improve their engineering properties. Source: US Dept. of Transportation

There are five principle reasons to compact soil:-

- Increases load-bearing capacity-

- Prevents soil settlement and frost damage-

- Provides stability-

- Reduces water seepage, swelling and contraction-

- Reduces settling of soil-source Soil Compaction Handbook.

Porosity

Bulk Density is an indirect measure of soil pore space. In fact, there is a formula: 100-((B.D. ÷ Particle Density) X 100)=% pore space or POROSITY. Under field conditions, pore spaces are occupied at all times by air and water. "Tortuous pathways" best describes soil pores. Porosity, when expressed as a percent, is the same thing as percent pore space. Soil particles have irregular shapes, and thus the spaces or pores between them vary irregularly in size, shape and direction. Sandy soils have large continuous pores, while clays have small pores which transmit water slowly. Clays, however, contain more pore space than sandy soils, because of the pores inside of the soil peds. To growing plants, pore sizes are of more importance than total pore space. We will be discussing this more in the unit on soil water.

Particle Density

In the laboratory you will determine porosity and particle density for the sandy soil used previously. Particle density will also be determined for this soil. Particle density only takes into account the volume occupied by the solid particles. It excludes the volume occupied by air and water. Since a large portion of most soils is composed of particles derived from quartz minerals, the particle density of most soils is near 2.6 g/cc, which is the density of quartz. Variations in the particle density are due to the presence of heavier minerals like iron oxides or lighter organic components.

Significance of Porosity

Above, a weakly developed soil is on the left and a well developed soil is on the right. Where clay accumulation is minimal, the macro pore space is about the same with increasing depth. However, where clay has moved into pores, the amount of macro pores and the rate of water movement both decrease. Soils with clay films will have slow water flow through the Bt horizon.

Some soils are difficult to aerate by plowing due to the turf cover. For lawns, playgrounds, or football fields, aeration is accomplished by making small holes in the turf with solid tines or using hollow tines to pull out a small core. Later the holes can filled with sand to promote air exchange. This is called *aerifying* and *topdressing*.

Many home lawns are compacted because of the construction activities when the home was built or due to people walking or riding on the turf when the soil is wet. Aerification will improve the turf growth on these lawns.

Aeration Equipment=Tines & John Deere

Soil Temperature Regimes

The temperature of a given soil at a given time is dependent upon the gains and losses of heat energy. Generally, dark surfaces will absorb more heat than light surfaces. However, the amount of water in the soil is an important factor. Losses of the absorbed heat are by radiation back into the atmosphere as long-wave radiation, heating the air and cooling the soil.

These soils warm quickly in the spring due to their dark surface.

Soil temperature regimes are used to classify soils. They are defined according to the average annual soil temperature in the root zone. The use of soils for agriculture and forestry is closely related to soil temperature, due to the specific requirements of plants.

Over most of the earth, daily soil temperatures below 50 cm deep seldom change. To approximate the mean annual soil temperature, 2 ° F is added to the mean annual air temperature. If we do this for the annual temperature map of Minnesota and follow the 45°F line, which would be 47° soil temperature, the state is divided along a line from the Twin Cities to the South Dakota border, and soils are frigid to the north and mesic to the south.

Chapter 2

Soil Sampling and Sample Preparation

Nature of Soil Variation

The starting point of a soil analysis is obtaining a representative soil sample. Obtaining this sample is not simple because soil is not just a neatly mixed variety of minerals, deposited in a uniform manner, but a complex living media of a very heterogeneous nature. Soil analysis is only as good as the soil sample used. When laboratory data is criticized, the fault has often occurred at the sample gathering stage.

Soil sampling methods vary depending on the reasons for the analyses:

- For analysis of soil fertility, the sample might be extracted from a depth comparable to the rooting zone
- For soil classification for mapping purposes, soils should be mapped and described according to their texture, colour and the thickness of each layer or horizon
- For non-agricultural purposes, parameters such as compaction, bearing capacity, shearing strength and infiltration are examined.

Soil is never homogeneous. Heterogeneity exist even among soils that are classified as belonging to the same unit. It is important to consult reference material on the subject and become familiar with a specific soil series before attempting to take a sample. Many factors may cause variations among soils. Some more influential factors are:

- vegetation
- topography
- cropping and tillage practices.

Vegetation. The type of plant life which covers a certain area of land can exert its influence in several ways. Changes in soil composition can occur when plant residue is incorporated into the soil. This is often the case after harvesting. Also, the structure of the soil is affected. The crop canopy and rooting system protect the soil from rain damage and severe temperatures, and make soil less susceptible to erosion and leaching.

Topography: Good top soil is often eroded from hilltops and ridges and can be found deposited, along with clay, in the lower areas. Figure 1 shows the influence of topography and cropping practices on soil acidity (pH). Each field has been planted with a particular crop for several years.

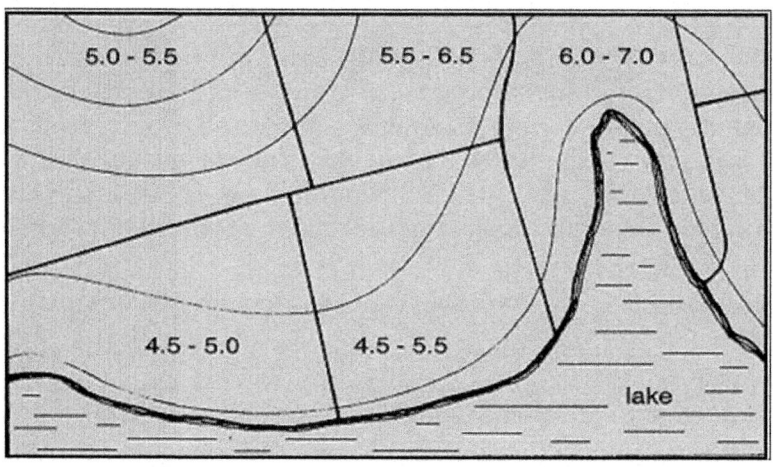

Figure 1. Soil acidity (pH) of different fields.

Cropping and tillage practices: The distribution of soil nutrients, and thereby the composition of soils, vary among those that are heavily tilled and those that are worked only slightly or not at all. As a result, samples from fields with different cultivation practices are to be taken from different depths.

Regardless of which factors have been most instrumental in causing variation in soils, whether variations are caused by vegetation, topography, cropping, tillage, or a combination of all four, they influence the intensity with which a soil must be sampled. The more heterogeneous the soils which make up a sample population, the more intense the sampling must be in order to obtain a given accuracy.

The first step in gathering soil samples is to make a detailed plan of the sampling technique used. This should include proper instructions on:

- the depth of sampling the soil
- the amount of sample required
- the number of samples required for a given accuracy

Judgment Sampling

Researchers, scientists, or farm managers may be called in when a crop shows a certain growing pattern or when surface differences are observed for a soil. For example, differences may occur in soil colour which may be the result of many factors. The researcher judges the colour differences: e.g. he may judge a particular shade of colour to be typical for a sample at certain sites. Then from these sites, samples are drawn. The accuracy of these samples depends totally on the judgment of the researcher-which may or may not be good. Some persons, in order to include as many extremes as possible, commit the error of over sampling. Probably less desirable is the person who takes the opposite approach, excluding the extremes and ending up with a sample which is not representative. In either case, the judgment of the sampler determines the accuracy of the results.

In certain situations, sample choices based on judgment are accurate enough. For example, if small sites are involved and no estimate of accuracy is needed, judgment sampling might be satisfactory. As the sample site becomes larger, and the selection of representative samples becomes more difficult and time consuming, judgment sampling is inaccurate and other sampling methods must be used.

Simple Random Sampling

Simple random sampling is a more precise method of taking soil samples and is less biased by the sampler than judgment sampling. Judgment sampling is used where soil or cropping differences are noticeable and where the focus of the survey is only on one particular area of the field.

Random sampling is needed where the soil differences are not immediately noticeable by colour, texture, etc. and selection of representative samples becomes more difficult. Simple random sampling could be used, for example, to take a soil sample in a small experimental plot where high accuracy is needed. In this case, sampler bias must be eliminated as much as possible for the sample to be representative of the entire plot. Following are the steps needed to take a random soil sample:

- Obtain a map or sketch of the area to be sampled
- Decide upon a scale for the sketch: if it is a large field, you may decide that 1 cm equals 100 m; if a small field, 1 cm could equal 1 m. (Use any convenient unit of measurement such as feet, paces, yards, etc. if preferred)
- Select a corner of this map and draw two lines connected at one point at right angles to each other. They do not need to be equal in length. Each will extend from the point at which they meet to the farthest point at the edge of the field. The important thing is to cover as much of the field as possible using two lines joined at one end. These lines are usually called the "axes"
- Using the map scale units you decided on, determine the length of the two axes. (Example: one axes is 28 cm and the other is 45 cm, corresponding to 28 m and 45 m in the field)
- Number slips of paper from 1 to the highest number of units on either of the two axes (Example: numbers 1 to 45). Place the numbers in a box and shake well
- Draw any one number between 1 and 45 from the box at random without looking (for example 27). Return the number to the box and draw a second number between 1 and 28 (for example 15)
- Plot the point of the first number on its axes and the other number on the other axes. These two points are called "coordinates"
- Plot the point where the coordinates intersect in the field. This is the approximate location where the sample will be taken
- Draw two new axes parallel to the first, at right angles, starting from the point you just marked
- Roughly measure the width of your sampling device. (Example: if you are using an auger, the auger might have a diameter of 5 cm)
- Divide the unit of measurement you decided on for the field by the diameter of the sampling device. (Example: We chose 1 m, which is 100 cm; 100 cm/5 cm = 20
- Put slips of paper numbered 1 to the number you just calculated into a box and shake. (Example: papers numbered 1 to 20)
- Draw two slips of paper. These are your new coordinates.

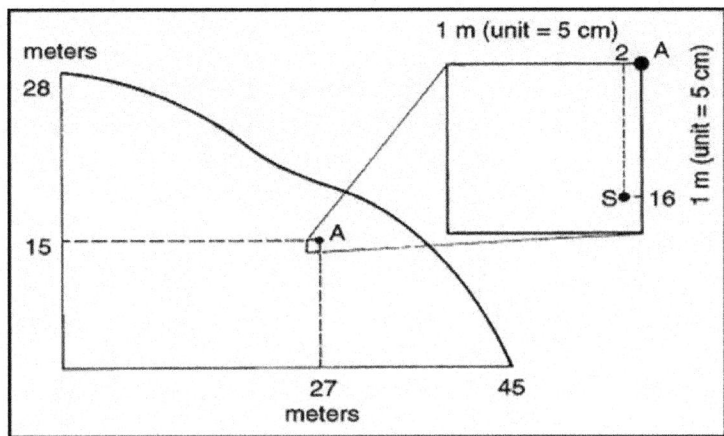

Figure 2. Random sampling.

- Find the spot in the field that corresponds to the intersection of the coordinates (point S). This is where you take your sample (Figure 2)

- Using the field units chosen, measure out in the field what you have plotted on paper, and take the sample with your sampling device.

Stratified Random Sampling

Farms or fields should be sampled according to soil areas. Different histories of crop management require separate samples even on the same soil. Eroded areas or other poor yielding spots must be sampled and evaluated separately. Samples have to be taken from each soil and crop sequence (Figure 3).

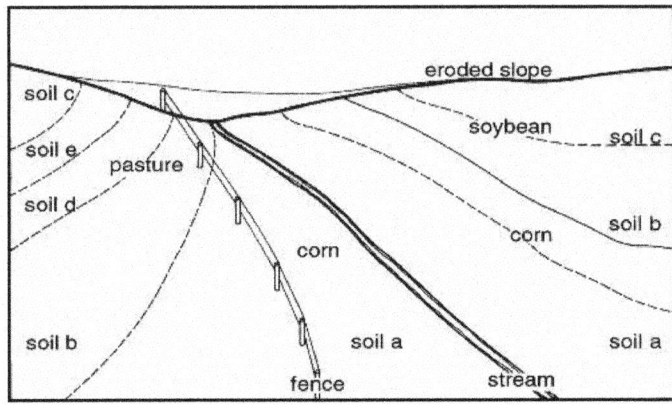

Figure 3. Stratification of soils (Smith 1976, modified).

When stratified random sampling is used, the population, i.e. a field or plot, is divided into sub-populations (strata). For example, the strata may be a ridge, a slope, a low area or a level area. A simple random sample is taken from each strata. The scientist or researcher prefers this sampling method for two reasons:

- to make a statement about the sub-population,

- to increase the accuracy of estimates over the entire population.

Since the sampler selects stratified random sampling for its accuracy, stratification must eliminate some variations caused by any sampling errors. In general, accuracy increases as stratification becomes more well defined. To estimate the sampling error of a stratum, sample at least two units. If the stratum is kept small enough, a good estimate of error can be obtained. One final point to keep in mind when applying this sampling technique is that the stratum is sampled in proportion to the total. If it is 20% of the total, then the number of sampling units to take from the stratum should be 20%.

Systematic Sampling

Systematic sampling is popular because it is accurate and relatively easy to use. Systematic samples are taken from sites that are equidistant from each other, either in one or two dimensions, forming a grid. Select the first unit at random. Take the following samples at uniform intervals.

Figure 4a consists of a grid formed when two sets of equidistant parallel lines are intersected at right angles to each other. Figure 4b consists of equidistant, parallel lines, set at 60 degree angles. Triangles are formed by drawing horizontal lines through the intersections.

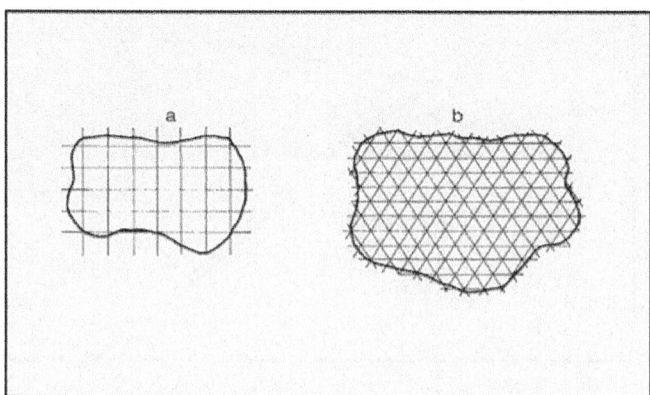

Figure 4. Grid patterns for systematic sampling.

Although the systematic grid pattern usually allows more accurate sampling, under certain conditions, it is inefficient. Madow (1944) found this to be the case when a fertility gradient existed, either along the rows or columns of a field. For this reason, systematic designs cannot be applied to fields having a slope or drainage problem without first considering the form of population distribution.

The main problem of systematic sampling is how to estimate sample error. There are several approaches:

- Assume that the population was in random order before the systematic sample was drawn. Estimate the sample error in the same way as the simple random sample.
- Block or stratify the sample, assuming that variations among units within a block are sample variations. Estimate the sample error as with the stratified random sample.
- Take a number of separate, systematic samples, drawn at random from all possible systematic samples of the same type. Calculate the mean and treat as a simple random sample.

Sources of Error

In the process of gathering soil samples for use in laboratory investigation, errors can occur at several stages:

- sampling errors
- selection errors
- measurement errors.

Sampling errors. Sampling errors cannot be eliminated entirely. Since sampling means to take a part, errors can be reduced by careful selection. Selection errors. Selection errors can occur when a sampler, eager to do a good job, over-samples the borders of a field. Sampling in rocky areas can also cause error. The errors often cancel each other. Otherwise the sampler may employ the two-step procedure of the simple random sampling which is designed to eliminate selection errors. Measurement errors. These types of errors arise in various circumstances:

- When the measurement taken is not the true value of the unit. One example is the random error. A random error takes place when cores of soil are mistakenly assigned constant weights, even though they are variable
- As a result of variations in analytical techniques. Fortunately, random errors of measurement tend to cancel each other as the sample size becomes larger

- Bias errors arise either because tare weights are ignored or because of an offset calibration of the appropriate curve

Sub-sampling and Composite Sampling

Subsampling. In many soil investigations, subsampling or multi-stage sampling is advantageous. To subsample, first divide a sample unit into small portions. Then choose a second sample from these portions and measure it according to the characteristics under consideration. This method of sampling, frequently used when bulk density cores are taken, saves both time and money. Composite sampling. When conducting laboratory analysis, using a composite sample instead of individual samples saves time and money. To obtain composite samples, first take a number of field samples of equal amount sufficient to represent the population. Then mix them to form one composite. Laboratory analyses is done on this composite or a subsample of it.

Sampling Tools and Sample Preparation

A sampling tool should be:

- uncontaminated
- approximately uniform in cross section to the desired depth
- provide reproducible sampling units.

Tapered cores or slices may bias the analysis result if systematic variations with depth are significant. Commonly used sampling tools are (Figure 5):

- blades: trowel, spade, shovel, spoon, knife, cutlass
- tubes: open-sided and plain cylinders
- augers: wood-bit, post-hole, sheathed auger.

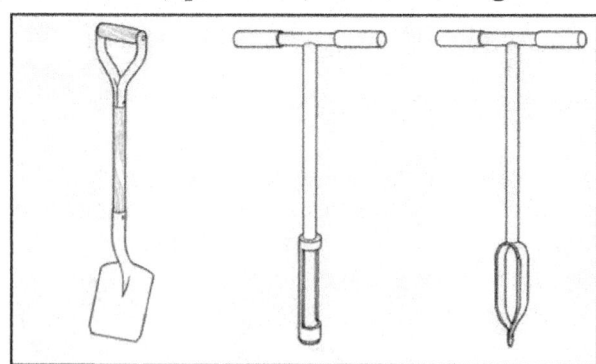

Figure 5. Blade, tube, and auger (left to right).

For comparison over periods of time, take soil samples at approximately the same time of the year (e.g. before planting). When making comparisons, consider other factors like weather conditions, crops, treatments, seasonal fluctuations.

Select sampling depth according to the purpose of sampling. For soil fertility evaluations in annual crops, sample at a depth of 0 to 15 or 0 to 20 cm. For perennial crops (e.g. trees), take deeper samples since tree roots often grow deep into the soil.

Soil samples usually need preparation before laboratory analysis:

- *Air-drying:* Crush large soil clods to facilitate drying. Do not dry at high temperature. During air-drying, avoid contamination (i.e. from dust, gases, rain, etc.). Air-drying usually takes one week

- *Crushing:* Crush the sample in a mortar using a rubber or porcelain-capped pestle (Figure 6). The mortar is usually made from porcelain

- *Pulverizing:* Not all samples require pulverizing. The subsampling error is a function of the ratio between the average weight of the largest particles and the weight of the subsample. If the subsample being analysed is small (i.e. for total nitrogen and organic carbon analysis), the sample has to be pulverized to a fine powder (less than 0.5 mm)

- *Sieving:* Sieve the soil through a 2 mm sieve made of brass, stainless steel, or plastic. Use plastic sieves when micronutrients are to be analysed

- *Mixing and Storage:* Mix samples thoroughly, then store in clean closed containers (i.e. polyethylene bags or bottles). Label the containers

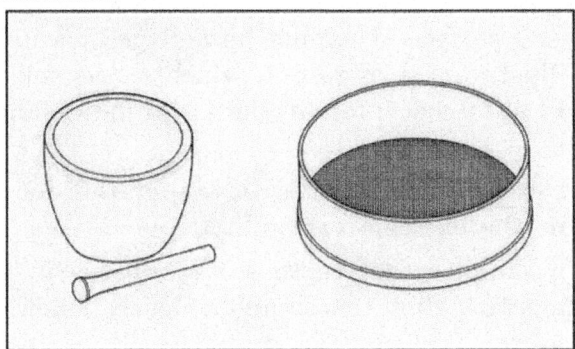

Figure 6. Mortar and pestle, sieve.

Sampling and Preparation

One of the most important aspects of analytical work is the proper collection and preparation of representative samples. The following are some general guidelines on sampling techniques for soils, plants and water.

Specific sampling information can be obtained from a variety of sources, including literature and Extension Specialists. This document also includes recommendations for preparing samples prior to submitting them to the ANR Analytical Lab.

If you have any questions regarding sampling or sample preparation, please contact the Lab at 530-752-0147.

Orchard Soil Sampling

Orchard soil testing has special value in the following situations.

1. Before planting or replanting—for prediction of fertilizer needs and/or of possible soil-related problems.
2. In established orchards—for prediction of fertilizer needs for the cover crop and, in some instances, for the trees.
3. In established orchards—for diagnosis of soil-related problems involving poor tree performance.

The Soil Variability Problem

Soils are variable. In fact, most surface soils vary a great deal within short distances across the landscape. The variability is much greater than most people realize. Some of the variability can be seen or anticipated because of obvious differences in slope, depth, texture, etc.

However, much of the variability is not visible, either because it is below the soil surface or cannot be detected except by soil tests. This is well illustrated in Figure 1, which shows soil test levels for phosphorus (P) at 50-foot intervals on a grid in a field that appeared to be uniform.

The field was heavily levelled in preparation for rill irrigation, which explains the extreme variability.

To obtain samples that represent conditions in the field, it is extremely important that the sampler closely follow the sampling instructions given.

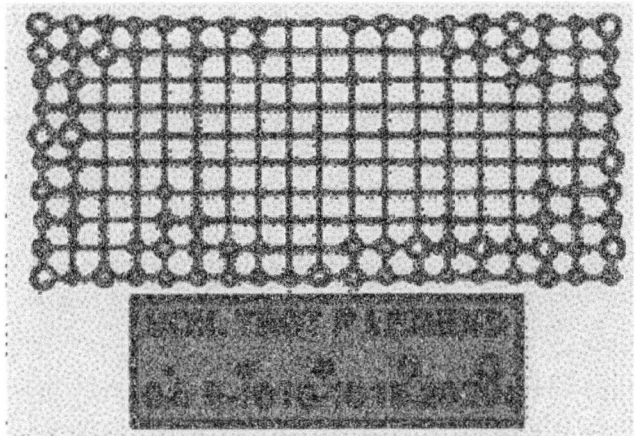

Figure 7. Variability in soil test P in a 12-acre field.

Guide to Proper Sampling

Sampling Tools

The recommended and most frequently used tool is the open-face, 36-inch soil sampling tube graduated to either 6 or 12 inches. The inside diameter is usually 3/4 inch and the open-face slot is usually 1 2 inches long (see Fig. 8).

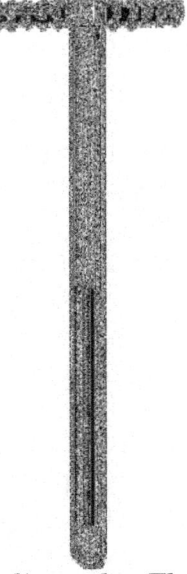

Figure 8. An open-face sampling tube. The tip is slightly enlarged on the outside and slightly tapered inward on the inside for efficient extraction of samples.

If a sampling tube is not available, one can use an irrigation shovel in an attempt to simulate the sampling done by a tube, but, at best, the sampling job will not be as good. If sampling is difficult because of gravel, hardpan, etc., one may have to use a soil auger. For sampling below 3 feet, a King tube is best in soils not having gravel or stones. The King tube requires pounding with a heavy hammer.

Sampling equipment should be of stainless steel. Containers should be plastic or plastic-coated to prevent contamination.

Before Planting

An established orchard involves a large capital investment and is expected to remain in production for many years. Improper or insufficient sampling usually means improper fertilization. Improper fertilization before planting cannot be readily corrected after planting and the problem may continue for the life of the orchard resulting in cumulative reduction in profits.

Considerable evidence indicates the advantages of applying such needed immobile nutrients, as potassium (K) or zinc (Zn) before planting long-term perennials, such as hops, grapes, or tree fruits.

1. Divide each field into sampling units. This refers to areas within a field that are known to be different from one another because of slope, soil depth, cropping and fertilizer history, drainage, areas of poor growth, etc. Each of such areas can be considered a sampling unit. If a sampling unit is less than 2 acres, take 8 to 10 soil cores, 1 foot in depth, throughout the unit. This is called a composite sample. If the unit is greater than 2 acres, sample it intensively.

2. Sample intensively (Fig. 3). In addition to the obvious variations are the hidden variations that nearly always occur in each field or sampling unit even though it may appear to be uniform.

    One composite sample from a large unit can be very misleading. Intensive sampling is an important new concept which considers soil variability and provides a basis for precision fertilization. Take at least one point sample from each acre. A point sample is a small composite of five cores within a 10-foot radius. In any case, never take only one point or one composite sample from fields larger than 2 or 3 acres.

3. Map the field. Obtain a Soil Conservation Service map of the farm showing soil type and physical properties, such as slope,

soil depth, and soil texture. Become familiar with the names and properties of all soils on the farm. This is an excellent resource and a useful base from which to map in further detail.

Proper sampling provides an excellent opportunity for mapping the field according to physical properties. This information is a valuable basis for making management decisions. One already has a good start when he has delineated the field into sampling units—areas having visible differences, such as slope, etc. Then during the process of intensive sampling, the sampling site in each acre can be examined with the soil tube to 3 feet for soil depth, texture, and possible physical problems. Thus, a map can be constructed based on visible delineations on the surface, plus information obtained by examination of the soil beneath the surface. This information can then be correlated with the original SCS map.

4. Sample to the proper depth. In general, soil test correlation research is based on the surface foot of soil. Therefore, take 0- to 1 2-inch samples when testing for pH, organic matter, salts, P, K, boron (B), and Zn. When testing for nitrate (NO_3N) before planting the orchard, sample by foot-depth increments to 3 feet.

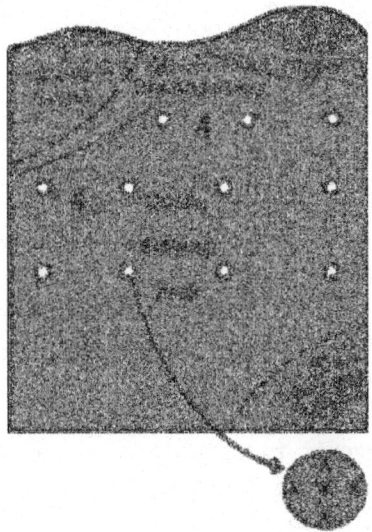

Figure 9. Soil sampling an 18-acre field containing five sampling units. Units 1, 2, 3, and 5, each comprising 1 acre or less, are sampled by taking one composite sample in each. Unit 4 is sampled

intensively. Each circle, approximately 10 feet in diameter, represents a "point" sample. Each x represents one sample core taken with a soil tube. Each point sample comprising five cores is kept separate from all others and is analysed separately.

Before Replanting

Soil sampling and testing procedures for orchards to be replanted can follow the same general principles as before a new planting except that:

1. Old apple tree sites should be sampled separately, if this is feasible. Tree sites can be identified either by sampling immediately after removal of the tree or by a very intensive sampling for short distances in two directions to determine areas of highest and lowest arsenic (As) content.

2. It may be advisable to run the special tests (As, salts, pH, B, NO_3N) as well as the tests for P, K, and Zn.

Established Orchards

Cover Crop. Soil testing for predicting fertilizer needs for established fruit trees is more complex than for field crops and has some limitations. However, except for N. It can be said that "what is good for the cover crop or sod is good for the trees." Soil testing of surface-foot samples taken between trees is considered to be of value for both cover crop and trees.

The sampler should follow somewhat the sampling principles outlined under "Before Planting" section. Such samples should be tested for pH, salts, P, EC, Zn, and B. Special Problems. In established orchards, soil chemical problems sometimes occur which relate to past management practices. The most common problem has to do with the placement of ammonium fertilizer in a circular band inside the drip line resulting in low pH and possible toxic levels of manganese (Mn). Other problems involve excessive levels of As, B, or salts. Sampling should be done where the fertilizer was applied—usually from at least halfway inside the drip line to the trunk.

A good procedure is to sample from good and poor trees for comparison. Tests should include pH, B. As, salts, and NO_3N. Interpretation should be made with the help of the Extension agent or other qualified person.

Monitoring Sites. Test soil every three years to monitor the levels of salts, pH, and B in the area of fertilizer application inside the drip line. Take samples from individual trees by foot-depth increments to 3 feet and from four sides of the tree. It would be desirable to select one tree in every acre for this purpose. The purpose is to avoid low pH levels, excessive salts, and either too low or too high soil B levels.

Soil Testing

Send samples to commercial laboratories located in various parts of the state. Assistance in sampling, packaging, and locating laboratories can be obtained through the county Extension offices.

Chapter 3

The Soil Foodweb: It's Importance in Ecosystem Health

Soil Foodweb Significance

The structure and function of the soil foodweb has been suggested as a prime indicator of ecosystem health. Measurement of disrupted soil processes, decreased bacterial or fungal activity, decreased fungal or bacterial biomass, changes in the ratio of fungal to bacterial biomass relative to expected ratios for particular ecosystems, decreases in the number or diversity of protozoa, and a change in nematode numbers, nematode community structure or maturity index, can serve to indicate a problem long before the natural vegetation is lost or human health problems occur.

Soil ecology has just begun to identify the importance of understanding soil foodweb structure and how it can control plant vegetation, and how, in turn, plant community structure affects soil organic matter quality, root exudates and therefore, alters soil foodweb structure. Since this field is relatively new, not all the relationships have been explored, nor is the fine-tuning within ecosystems well understood.

Regardless, some relationships between ecosystem productivity, soil organisms, soil foodweb structure and plant community structure and dynamics are known, and can be extremely important determinants of ecosystem processes. Alteration of the soil foodweb structure can result in sites which cannot be regenerated to conifers, even with 20 years of regeneration efforts. Work in intensely disturbed forested ecosystems suggests that alteration of soil foodweb structure can alter the direction of succession. By managing foodweb structure appropriately, early stages of succession can be prolonged, or deleted. Initial data indicates that replacement of grassland with forest in normal successional sequences requires alteration of soil foodweb structure from a bacterial-dominated foodweb in grasslands to a fungal-dominated foodweb in forests.

In addition to responses to disturbance, it is clear that species diversity, community diversity and foodweb complexity increases with increasing successional stage. Indeed, examination of foodweb interactions and ecosystem diversity, instead of community diversity, may result in new ecosystem measures which reflect this increased community diversity and increased connectivity in later successional stages.

The numbers, biomass, activity and community structure of the organisms which comprise the soil foodweb can be used as indicators of ecosystem health because these organisms perform critical processes and functions. Soil decomposers (bacteria, fungi and possibly certain arthropods) are responsible for nutrient retention in soil. If nutrients are not retained within an ecosystem, future productivity of the ecosystem will be reduced as well as cause problems for systems into which those nutrients move, especially aquatic portions of the landscape.

As ecosystems become more productive, the total amount of nutrients retained within the system increases. As succession occurs, nutrients are increasingly immobilized in forms that are less available for plants and animals, such as phytates, lignins, tannins, humid and fulvic acids. In order for nutrients to become available once again to plants and animals, they must be mineralized by the interaction of decomposers, i.e. bacteria and fungi, and their predators, i.e. protozoa, nematodes, microarthropods, and earthworms (if present). These predator populations and the rates at which they perform mineralization processes are important to ecosystem stability. The activity of these predator-prey interactions (which determines the rate at which mineralization occurs) are in turn affected, and perhaps controlled by, higher level predators such as millipedes, centipedes, beetles, spiders, and small mammals.

It is perhaps something of a conundrum that in healthy ecosystems, while nutrient cycling and productivity increases, nutrient loss is minimized. What makes this possible is the increasing complexity of the soil foodweb. As total ecosystem productivity increases, biodiversity below ground, i.e., the structure and function of the soil foodweb, also increases. The greater the foodweb complexity, i.e., the interaction of decomposers, their predators, and the predators of those predators responsible for nutrient cycling and the retention of nutrients within the soil , the fewer the losses of nutrients from that system, the more tightly nutrients cycle from retained forms to plants, and back again. Without the soil foodweb, plants would not obtain the nutrients necessary for growth, and the above ground foodweb would not long continue.

Interactions of decomposers with their predator groups (protozoa, nematodes and microarthropods) maintain normal nutrient cycling processes in all ecosystems. Plant growth is dependent on microbial nutrient immobilization and soil foodweb interactions to mineralize nutrients. In undisturbed ecosystems, the processes of immobilization and mineralization are tightly coupled to plant growth. Following disturbance, this coupling is lost or reduced.

By monitoring soil organism dynamics, we can detect detrimental ecosystem changes and possibly prevent further degradation. The response of each group of soil organisms, i.e., soil saprophytic bacteria, symbiotic bacteria, saprophytic fungi, mycorrhizal fungi, protozoa, and nematodes, with respect to their total biomass and activity, can be used to indicate effects of contaminants on soil health. Instead of relying on an indirect measure of whether total biomass or activity is reduced , active and total biomass of each organism group can be directly measured. Lal and Stewart (1992) reviewed the relationship between system health and soil organic matter, and suggested that soil organism losses correlate with detrimental ecosystem changes. Development of the relationship between soil foodweb structure and function and assessment of potential toxic impact could be extremely useful for assessing ecosystem health.

The ratio of fungal to bacterial biomass and the Maturity Index for nematodes (Bongers, 1985). Both appear to be useful predictors of ecosystem health, although they must be properly interpreted given the successional stage being examined. For example, recently disturbed systems have nematode community structures skewed towards opportunistic species and genera, while the less opportunistic, more K-selected species of nematodes return as time since-disturbance increases. Thus, healthier soils tend to have more mature nematode community structures. However, as systems mature, nutrients tend to be more sequestered in soil biomass and organic matter, and thus the maturity index reflects an optimal, intermediate disturbance period in which greatest ecosystem productivity is likely to occur.

Ratios of fungal to bacterial biomass also predict this type of response. Highly productive agricultural soils tends to have ratios near one, but as a system undergoes succession into a grassland, this ratio dips downwards, indicating that for a healthy grassland system, the ratio should be less than one. In other words, bacterial-biomass dominates in healthy grassland soils. However, as succession proceeds yet further, fungal biomass begins to dominate and healthy forest systems have fungal to bacterial biomass ratios of greater than one, usually greater than 10.

Piparian or deciduous forests appear to be intermediate within this range of values. Alder forest soils are dominated by bacterial biomass, while popular forest soils are fungal-dominated. Clearly, further investigation is required.

The predators of bacteria and fungi tend to follow the dominance of the decomposer groups. Thus, bacterial-dominated soils have a majority of bacterial-predators (protozoa and bacterial-feeding nematodes), while fungal-dominated soils have a majority of fungal predators (fungal-feeding nematodes and fungal-feeding microarthropods).

Much work is still required at the bacterial and fungal species level. While the species of protozoa and nematodes have been researched in soils of this area of the west, publication of much of this information has yet to occur. Updates will be required as this information becomes available.

The Soil Foodweb: Function

Bacteria and fungi perform one of the major nutrient cycling processes, nutrient retention, in soil. The amount of N. P. S and other nutrients immobilized in bacterial and fungal biomass can be considerable, from several micrograms to milligrams of biomass, comprising a significant portion of the stable nutrient pool. When the bacterial or fungal component of the soil declines, more nutrients are lost into the ground and surface water. A major means of retaining nutrients may also be arthropod fecal material , depending on the ecosystem.

Soil bacteria are important in maintaining normal nutrient immobilization and decomposition processes in all ecosystems. Plants are strongly influenced by the presence of bacteria in the rhizosphere, especially with respect to microbial immobilization of nutrient, and mineralization of nutrients from bacterial biomass by predators. Disturbance of these soil processes may result in the un-coupling of mineralization and plant growth, with the resultant loss of nutrients from the system, causing problems for systems into which nutrients move.

As climate changes occur, bacterial populations in the soil could be significantly impacted. As temperature increases, bacterial numbers could increase, resulting in greater immobilization of nutrients in their biomass, causing greater nitrogen limitation of plant growth. Alternatively, bacteria could be inhibited by increases in carbon dioxide, resulting in decreased decomposition of soil organic matter and plant litter, which ultimately would change soil structure and nutrient cycling.

In addition, current work indicates that alterations in the fungal to bacterial biomass ratio strongly impacts vegetative community structure. If a forest soil, usually strongly dominated by fungi, loses the fungal component, reflected by a decrease in the ratio of fungi to bacteria, conifer species may be at risk of death. If the fungal to bacterial biomass ratio decreases past one, re-establishment of conifer species may be impossible.

Saprophytic fungi and bacteria form the base of the detrital foodweb, and as such are critically important for supporting the nutrient cycling subsystem of any ecosystem, landscape, or biome. Bacterial and fungal pathogens of plants, insects, rodents, and other organisms can control the population density of their hosts.

Mutualist bacteria and fungi can be critically important for plants and animals alike, for example, nitrogen-fixing bacteria on legumes, or rumen bacteria in cows, deer or elk. Without their mutualists, these plants and animals are not capable of competing with other organisms and become locally extinct. While methods are not yet capable of distinguishing between saprophytic and pathogenic species of bacteria and fungi in soil, their total and active biomass, and effects of different disturbances on their distributions, can be estimated. However, work should continue on methods to differentiate bacterial and fungal community composition in soil.

Protozoa, comprised of the three groups; (1) flagellates, (2) amoebae (both naked and testate), and (3) ciliated, are important in maintaining plant-available N and mineralization processes and, as bacterial-feeders, are important in controlling bacterial numbers and community structure in the soil (Foissner 1986). The presence or absence of certain protozoa species is indicative of the presence of certain hazardous wastes and therefore may be highly useful indicator organisms of certain types of environmental impacts.

Nematodes are one of the most ecologically diverse groups of animals on earth, existing in nearly every habitat. Nematodes eat bacteria, fungi, algae, yeasts, diatoms and may be predators of several small invertebrate animals, including other nematodes. In addition, they may be parasites of invertebrates, vertebrates (including man) and all above and below ground portions of plants. Nematodes range in length from 82 um (marine) to 9 m (whale parasite) but most species in soil are between 0.25 and 5.5 mm long.

Nematodes are recognized as a major consumer group in soils, generally grouped into four to five trophic categories based on the nature of their food, the structure of the stoma and esophagus and method of feeding. Plant-feeding nematodes possess stylets with a wide diversity of size and structure and are the most extensively studied group of soil nematodes because of their ability to cause plant disease and reduce crop yield. Fungal-feeding nematodes have slender stylets but are often difficult to categorize and have been included with plant-feeders in many ecological studies. Bacterialfeeding nematodes are a diverse group and usually have a simple stoma in the form of a cylindrical or triangular tube, terminating in a teeth valve-like apparatus which may bear minute nematodes (marine) to 9 m 0.25 and 5.5 mm

Predatory nematodes are usually large species possessing either a large styles or a wide cup-shaped cuticular-lined stoma armed with powerful teeth. Omnivores are sometimes considered as a fifth trophic category of soil nematodes. These nematodes may fit into one of the categories above but also ingest other food sources. For example, some bacterial feeders may also eat protozoa and/or algae and some stylet-bearing nematodes may pierce and suck algae as well as fungi and/or higher plants. Stages of animal parasitic nematodes, such as hookworms, may also be found in soils but generally are not common in most soil samples.

Nematodes and protozoa function as regulators of mineralization processes in soil . Bacterial-and fungal-feeding nematodes release a large percent of N when feeding on their prey groups and are thus responsible for much of the plant available N in the majority of soils. Nematode-feeding also selects for certain species of bacteria, fungi and nematodes and thereby influences soil structure, carbon utilization rates, and the types of substrates present in soil (Ingham, R. 1992). Root-feeding nematodes are among the greatest pests in agricultural systems and, with the loss of many nematicides, are becoming a greater concern. Without doubt, plant establishment, survival and successional processes are influenced by these soil organisms

Soil processes are important for maintaining normal nutrient cycling in all ecosystems. Plant growth is dependent on the microbial immobilization and soil foodweb interactions to mineralize nutrients. In undisturbed ecosystems, the processes of immobilization and mineralization are tightly coupled to plant growth but following disturbance, this coupling may be lost or reduced. Nutrients may be no longer retained within the system, causing problems for systems into which nutrients move. Measurement of disrupted processes may allow determination of a problem long before normal cycling processes are altered, before the natural vegetation is lost, or human health problems occur. By monitoring soil organism dynamics, we can perhaps detect detrimental ecosystem changes and possibly prevent further degradation.

Immobilization of nutrients in soil, i.e., retention of carbon, nitrogen, phosphorus, and many micronutrients in the horizons of soil from which plants obtain their nutrients, is a process performed by bacteria and fungi. Without these organisms present and functioning, nutrients are not retained by soil, and the ecosystem undergoes degradation. Thus, to assess the ability of an ecosystem to retain nutrients, the decomposed portion of the ecosystem, i.e., active and total fungal biomass, and active bacterial biomass must be assessed.

The Soil Foodweb: Structure

What is the soil foodweb? Per gram of healthy soil, which is about a teaspoon of soil plus organic matter, the following organisms are found: of which are mostly unknown to scientists. Bacteria break down easy to-use organic material, and retain the nutrients, like N. P and S. in the soil. About 60% of the carbon in those organic materials are respired as carbon dioxide, but 40% of that carbon is retained as bacterial biomass. The waste products bacteria produce become soil organic matter. This "waste" material is more recalcitrant than the original plant material, but can be used by a large number of other soil organisms, exemplifying the classic statement that "One man's garbage is another's treasure". Productive garden soil should contain more bacteria than any other kind of organism, although care must be taken to make sure beneficial bacteria, instead of disease-causing bacteria, are most prevalent.-S to 60 000 meters of fungal hyphae. Fungi break down the more recalcitrant, or difficult-to-decompose, organic matter, and retain those nutrients in the soil as fungal biomass.

Just like bacteria, fungal waste products become soil organic matter, and these waste materials are used by other organisms. Gardens require some fungal biomass for greatest productivity, but in order for best crop growth, there should be an equal biomass of bacteria as compared to fungi. Most grasslands or pastures have less fungi than bacterial, while all conifer forests have much more fungal, as compared to bacterial, biomass. As with bacteria, some fungi cause disease and the soil must be managed to prevent these fungi from being a problem.

-100 to 100,000 protozoa. These organisms are one-celled, highly mobile organisms that feed on bacteria and on each other. Because protozoa require 5 to 10-fold less nitrogen than bacteria, N is released when a protozoan eats a bacterium. That released N is then available for plants to take up. Between 40 and 80% of the N in plants can come from the predator-prey interaction of protozoa with bacteria.

-5 to 500 beneficial nematodes. Beneficial nematodes eat bacteria, fungi, and other nematodes. Nematodes need even less nitrogen than protozoa, between 10 and 100 times less than a bacterium contains, or between 5 and 50 times less than a fungal hyphae contains. Thus when bacterial-or fungal-feeding nematodes eat bacteria or fungi, nitrogen is released, making that N available for plant growth. However, plant-feeding nematode are pests because they eat plant roots. These "bad" nematodes can be controlled bacteria, fungi, and other nematodes. Nematodes need even less nitrogen than protozoa, between 10 and 100 times less than a bacterium contains, or between 5 and 50 times less than a fungal hyphae contains.

Thus when bacterial-or fungal-feeding nematodes eat bacteria or fungi, nitrogen is released, making that N available for plant growth. However, plant-feeding nematode are pests because they eat plant roots. These "bad" nematodes can be controlled biologically, as they are in natural systems, by fungi that trap nematodes, by having fungi that colonize root systems and prevent nematode attack of roots, or by predation of nematodes by arthropods. In cases of extreme outbreaks, however, the only answer may be the use chemicals to control these plant-feeding nematodes. However, once a chemical is used which kills the beneficial nematodes as well as the plant-feeding ones, the beneficial nematodes need to be replaced through inoculation.

A few to several hundred thousand microarthropods. These organisms have several functions. They chew the plant leaf material, roots, stems and boles of trees into smaller pieces, making it easier for bacteria and fungi to find the food they like on the newly revealed surfaces. The "comminuting" arthropods can increase decomposition rates by 2-to 100- times, although if the bacteria or fungi are lacking, increased decomposition will not occur. In many cases, however, the arthropods carry around an inoculum of bacteria and fungi, making certain the food they want is inoculated onto the newly exposed surfaces! Arthropods then feed on bacteria and fungi, and because the C:N ratio of arthropods is 100 times greater than the bacteria and fungi, they release nitrogen which then is available for plant growth. Some arthropods eat pest insects, while others eat roots. Again, it's important to encourage the beneficial ones and discourage the ones that eat plants!

The Web of Life Can Be Degraded

The interactions between these organisms form a web of life, just like the web that biologists study above ground. What most people don't realize is that the above ground wouldn't exist without the below ground systems in place and functioning. Soil biology is understudied, compared to the above ground, yet it is important for the health of gardens, pastures, lawns, shrublands, and forests. If garden soil is healthy, there will be high numbers of bacteria and bacterial-feeding organisms. If the soil has received heavy treatments of pesticides, chemical fertilizers, soil fungicides or fumigants that kill these organisms, the tiny critters die, or the balance between the pathogens and beneficial organisms is upset, allowing the opportunist, disease-causing organisms to become problems.

Over-use of chemical fertilizers and pesticides have effects on soil organisms that are similar to over-using antibiotics. When we consider human use of antibiotics, these chemicals seemed a panacea at first, because they could control disease. But with continued use, resistant organisms developed, and other organisms that compete with the disease-causing organisms were lost. We found that antibiotics couldn't be used willy-nilly, that they must be used only when necessary, and that some effort must be made to replace the normal human-digestive system bacteria killed by the antibiotics.

Soils are similar, in that plants grown in soil where competing organisms have been knocked back with chemicals are more susceptible to disease-causing organisms. If the numbers of bacteria, fungi, protozoa, nematodes and arthropods are lower than they should be for a particular soil type, the soil's "digestive system" doesn't work properly. Decomposition will be low, nutrients will not be retained in the soil, and will not be cycled properly. Ultimately, nutrients will be lost through the groundwater or through erosion because organisms aren't present to hold the soil together.

The best way manage for a healthy microbial ecosystem in a home garden is to routinely apply organic material, such as compost. To keep garden soil healthy, the amount of organic matter added must be equal to what the bacteria and fungi use each year.

Indiscriminate use of chemical fertilizers and pesticides should be avoided. If the soil is healthy for the type of vegetation desired, there should be no reason to use pesticides, or fertilizers. If a decision is made to change from grass to garden, or forest to lawn, a massive change in the soil foodweb structure is required and chemical use, along with judicious addition of the right kind of compost with the right kinds of organisms, may be necessary for a few years. But once the correct soil foodweb structure is in place, there should be no reason to apply chemicals.

If both bacteria and fungi are lost, then the soil degrades, than any other organism. If bacteria are killed through pesticide or chemical applications, and especially if certain extremely important bacteria like nitrogen-fixing bacteria or nitrifying bacteria are killed, fungi can take over and crop production can be harmed. For example, current research indicates that the reason moss takes over in lawn ecosystems is because the soil is converted from a bacterial dominated system to one dominated by fungi nutrients are lost, erosion increases and plant yield is reduced. If inorganic fertilizers are used to replace the lost nitrogen, the immediate effect may be to improve plant growth. However, as time goes on, it is clear that inorganic fertilizers can't replace the other kinds of food that bacteria and fungi need. After awhile, fertilizer additions are a waste of money, because there aren't enough soil organisms to hold on to the nutrients added. Surface and groundwater will become contaminated with the lost nutrients, causing problems.

Maintaining and Enhancing the Soil Foodweb

Bacterial dominance is maintained by mixing plant material into the soil. But the bacteria and fungi eat this material at an amazingly rapid pace and new inputs are required every year; Fungi can be maintained by letting litter accumulate on the soil's surface. Larger soil organisms like millipedes, centipedes, earthworms, and ants mix plant material into soil and open air channels, especially important in wet periods in heavy clay soils. To maintain a one-to-one ratio of bacteria and fungi needed for crop systems, a balance is needed between too much and too little mixing. Plant material needs to be mixed in enough to maintain bacterial dominance, but too much mixing results in soil degradation. Timing of mixing is important as well, but the optimal combination hasn't been determined for soil organisms in different types of soil.

It's important to remember that grassland, garden and forest soils represent a gradient from bacterial to fungal dominance. Gardens require equal amounts of bacteria and fungi, while trees require fungi. There are a number of examples where the fungal component has been lost from forest soils and as a result, tree regeneration is impossible. If the soil foodweb was better understood, there would ways to fix the problem, but that research is yet to be done.

In order to determine the organisms in soil, the biomass and activity of bacteria and fungi, the numbers of protozoa and nematodes, the types (beneficial and root-feeding) of nematodes, and VA mycorrhizal colonization of roots need to be assessed. Reference information on the biomass, numbers and types of these organisms is being determined for soils all over the world. The goal is to determine what the healthy soil foodweb structure should be for every soil type, given vegetation and climate characteristics. If the foodweb structure is not at that healthy level, another goal is to determine what it will take to return it to a healthy level. Once a healthy foodweb structure is achieved, the only time testing would be needed is when some problem is detected, suggesting the foodweb has changed in an unproductive direction.

Soil Sampling

Soil sampling should result in three samples from any particular area, such as a meadow, crop field, forest stand or garden. Five samples per area, or more would be preferable, but time and cost of analysis must be a consideration. The idea is to take enough samples that the variability within that area can be assessed. One possible approach is to mentally split the area to be sampled into three equal areas. From each of the three areas, between three and ten small soil cores should be mixed together in a plastic bag. The cores should be taken by pushing aside the litter (loose recognizable plant litter material) on the top of the soil and removing soil (may contain some unrecognizable plant material, but is mostly mineral soil material, or sand, silt and clay fractions) from the 0-5 cm depth. The core should be about 2.5 cm or 1 inch diameter, and all the soil from this small cylinder should be removed and placed in the plastic bag. If mycorrhizal colonization is to be performed, the roots in each core should be removed and placed in the plastic bag. Small scissors should be used to cut the small roots.

In fact, the foodweb structure in any kind of material, from lake sediment, to rumen material from cattle can be assessed, but most research has been performed on soil-related material.

Interpretation of Soil Foodweb Structure

Ratio of Total Fungal to Total Bacterial Biomass

By examining the structure of the soil foodweb in a range of soils, all grassland and most agricultural soils have ratios of total fungal to total bacterial biomass less than one (F/B < 1). Another way to interpret this is that the bacterial biomass is greater than the fungal biomass in these soils.

In the most productive agricultural systems, however, the ratio of total fungal to total bacterial biomass equals one (F/B = 1) or the biomass of fungi and bacteria is even. When agricultural soils become fungal-dominated, productivity will be reduced, and in most cases, liming and mixing of the soil (plowing) is needed to return the system to a bacterial-dominated soil.

All conifer forest soils are fungal dominated, and the ratio in all forest soils in which seedling regeneration occurs is above 10. In general, productive forest soils have ratios greater than 100. This means that fungal biomass strongly outweighs the bacterial biomass in forest soils. In the case where forest soils lose this fungal-dominance, it is not possible to re-establish seedlings. When forest soil becomes bacterial-dominated, conifer seedlings are incapable of being re-established. In the few studies of riparian forests that have been performed, some deciduous riparian forest soils are bacterial-dominated. In the case of riparian aspen and beech soils, the soils are bacterial dominated. But poplar, oak and maple soils are is fungal-dominated, although not as strongly fungal-dominated as in conifer systems. No studies on establishment of seedlings in these systems have been performed.

The ratio of total fungal to total bacterial biomass has been related to ecosystem productivity, but numbers or length of active and total bacteria and fungi are also indicative of the health of soil. For different soils, vegetation and climate, the density of bacteria or fungi indicate the past degradation of the soil. As explained above, and again in the following sections, bacterial numbers should be greater than one million for all agricultural soils, preferably nearer 100 million for the most productive soils. For the most productive forest soils, for example, fungal length should be above 5000 meters of hyphae per gram soil.

Biomass of Total Fungi

Fungal biomass is extremely important in all soils as a means of retaining nutrients that plants need in the upper layers of the soil, i.e., in the root-zone. Without these organisms to take-up nutrients, and either retain those nutrients in their biomass, or to sequester those nutrients in soil organic matter, nutrients would wash through the soil and into ground or surface water. Plants would suffer from lack of nutrients cycling into forms that the roots can take-up, if these nutrients aren't first immobilized in the soil through the action of fungi or bacteria. For forest soils, fungi sequester most of the nutrients, although significant portions are immobilized by bacteria as well.

In soil in which only fungi are present, the soil will become more acidic, from secondary metabolites produced by fungi. Aggregates are larger in fungal-dominated soils than in bacterial-dominated soils, and the major form of N is ammonium, since fungi do not nitrify N. These conditions are more beneficial for certain shrubs, and most trees.

Total fungal biomass varies depending on soil type, vegetation, organic matter levels, recent pesticide use, soil disturbance and a variety of other factors, many of which have not been researched completely. However, for normal grassland soils, total fungal biomass levels are usually around 50 to 500 meters per gram of soil. For agricultural soils, fungal biomass is around 1 to 50 meters per gram soil, while for forest soils, fungal biomass is between 1000 meters to 60 km per gram of soil. More work is necessary to establish what the optimal fungal biomass value should be for each type of crop, soil, organic matter, climate, etc. Very little information is available for tropical systems, but that small amount of data indicates that temperate systems perform very differently from tropical soils.

The average diameter of hyphae in most soils is about 2.5 micrometers, indicating typical mixtures of zygomycetes, ascomycete and basidiomycetes species. On occasion the average diameter may be greater than 2.5 micrometers, indicating a greater than normal component of basidiomycete hyphae, while on other occasions, the average diameter of hyphae may be less than 2.5 micrometers, indicating a change in species composition of soil fungi to a greater proportion of lower fungi. Actinomycetes are not usually differentiated from fungi, since actinomycetes are hyphal in morphology and are rarely of significant biomass. In some agricultural soils, these narrow diameter "hyphae" are of considerable importance, as demonstrated by Dr. A. Van Bruggan.

Biomass of Active Fungi

Activity in all soil organisms follow a typical seasonal fluctuation. This cycle is related to optimal temperature and moisture, such that a peak in activity usually occurs in the spring as temperature and moisture become optimal after cold winter temperatures. In systems where snow accumulates on the Coil surface, such that the soil does not actually freeze, fungal activity may continue at high levels throughout the winter in litter. Decomposition may continue at the highest rates through the winter under the snow in the litter. In systems where moisture becomes limiting in the summer, activity may reach levels even lower than in the winter. When temperatures remain warm in the fall and rain begins again after a summer drought, such as in Mediterranean climates, a second peak of activity may be observed in the fall. If these peaks are not observed, this suggests inadequate organic matter in the soil.

Numbers of Total Bacteria

Just as fungi are the most important players in retaining nutrients in forest soil, bacteria are the important players in agricultural and grassland soils. Bacteria retain nutrients first in their biomass, and second, in their metabolic by-products. In soil in which only bacteria are inoculated, the soil will become more alkaline, will have small aggregates, and generally will have nitrate/nitrite as the dominant form of N. These conditions are beneficial for grasses and row crop plants.

Numbers of total bacteria generally remain the same regardless of soil type or vegetation. Total bacterial numbers range between 1 million and 100 million per gram soil in agricultural soils, and between 10 million and 1,000 million in forest soils. Bacterial numbers can be above 100 million in decomposing logs, in anaerobic soils, in soil amended with sewage sludge or in soil with high amounts of comported material. In some instances following pesticide treatment, bacterial numbers can fall below 1 million, and this has been correlated with signs of severe nitrogen deficiency in plants. Bacterial numbers can drop to extremely low levels, below 100,000 per gram of soil, in degraded soils where nutrient retention is a problem.

Biomass of Active Bacteria

As with active fungal biomass, bacterial activity usually peaks in the spring and decreases during the summer with drought. If the temperature remains warm in the fall and fall rains begin, a second peak of activity usually occurs. The ratio of active fungal to active bacterial biomass, even in forests, shows that bacterial biomass is usually more active than fungal biomass. If these peaks with temperature and moisture are not seen, then lack of appropriate food in the soil to support bacterial and fungal biomass is suspected. If bacterial and fungal activity does not respond to seasonal fluctuations, then subsequent impacts on the predators in the soil will be observed.

Protozoan Numbers

Protozoa feed on bacteria, and as they feed on their prey, N is released. It's unclear just how much N is released per individual feeding event, since it depends on whether the bacterium was actively growing, thus containing more N. or whether the bacterium was in a resting starving phase, and containing much less N. Several studies have shown that a major portion t40-80%) of the nitrogen that cycles through in certain agricultural soils is cycled by protozoa. Without these organisms in soil, plants suffer a significant reduction in available N. However the optimal relationship between the number of bacteria and the number of protozoa has not been quantified.

There appears to be a great range in protozoan numbers from soil to soil, and even from field to field. Some of the observations that have been made, when dealing with agricultural soil tie., bacterial dominated) is that when protozoan numbers are high, bacterial-feeding nematode numbers will be low, and vice versa. Thus there appears to be significant competition between bacterial-feeding predators for the bacterial prey. Whether this is indicative of the type of bacteria present in the soil, and whether this has any relationship to productivity in agricultural situations is not known.

Testate amoebae are only found in significant and constant numbers in forest soils, and are never found in temperate agricultural soils. Why this is the case is not known, but continues to be observed.

Nematode Numbers, Community Structure

There are four major types of nematodes, which includes bacterial-feeding, fungal-feeding, root-feeding and predatory nematodes. All nematodes are predators, and thus reflect to some extent the availability of their prey groups. However, other organisms prey upon these nematodes as well, and nematode numbers can also reflect the balance between the availability of nematode prey, as well as feeding by nematode predators.

Both bacterial-feeding and fungal-feeding nematodes mineralize N from their prey groups. Bacterial-feeding nematodes are more important in bacterial-dominated soils (agriculture and grassland systems), while fungal-feeding nematodes are more important in fungal dominated soils (conifer and most deciduous forests). Between 70 and 80% of the nitrogen in rapidly-growing trees has been shown to come from interactions between nematode predators and their prey. Between 30 and 50% of the N in crop plants appears to come form the interactions of bacterial-feeding nematodes and bacteria. Thus, the presence and numbers of bacterial- and fungal-feeding nematodes is extremely important for productive soils.

Root-feeding nematodes are detrimental to plant growth. As few as one endo-parasitic nematodes per plant may be enough to result in decreased productivity or death, while plants may tolerate several hundred ecto-parasitic nematodes per root system without reduction in production. Compensatory plant production has been observed with a little root-feeding, in that plant production is greater with a few herbivores munching on the plant than without feeding taking place.

Root-feeding nematode numbers can be reduced by competition for root space. VA mycorrhizal fungi may prevent root-feeding nematodes from reaching the roots through a variety of mechanisms. Nematode trapping fungi trap and kill many root-feeding nematodes. Other fungi and bacteria may be active inhibitors of nematode presence in the rhizosphere. Effective biocontrol of these plant-feeders is being worked on and may be possible in the near future.

T. Bongers, in the Netherlands, suggested the use of a Maturity Index for nematodes in soil. Certain species of nematodes are more commonly found following disturbance, while other species are more typical inhabitants of less-disturbed soils. The numbers of the four different trophic groups, or of different genera or species of matodes can be interpreted in several ways. First, significant differences in the number of individuals in a trophic group, or in nematode species in any treatment, compared to the control or reference treatment indicates an impact on that particular group or species. Several studies have recently shown that changes in numbers of even a single species of nematode in soil can significantly alter nutrient cycling within a soil. However, until more work is done to determine whether the same (or similar) nematode species control that same process in a similar way in other soils, extrapolation of results from a study performed in a different soil, with different plant species, different bacterial and fungal species, and different numbers of competing predators remains fraught with difficulty. However, it clearly suggests that alteration in nematode species composition could have negative impacts on plant growth, plant species composition, and thus ecosystem productivity.

A second way of interpreting nematode data is to assess changes in process rates. Since a bacterial-feeding nematode consumes 106 bacteria per day, reductions in this trophic group should result in an increase in bacterial biomass, with concomitant increase in net nutrient immobilization, and a decrease in nutrient mineralization with a concomitant effect on availability to plants. A fungal-feeding nematode consumes the cytoplasm in 10-50 meters of hyphal length per day, with similar effects on nutrient cycling as decreases in bacterial-feeding nematodes.

When root-feeding nematodes numbers are decreased, the ecosystem impact is positive, since root-feeding nematodes reduce plant growth/yield. However, root-feeding nematodes are highly opportunistic organisms, and are among the first organisms to invade after disturbance. Thus, one result of any disturbance which seriously affects ecosystem stability is reduction in the number of organisms which displace root-feeding nematodes. These competitors of root-feeding nematodes are mycorrhizal fungi, nematode-feeding nematodes, nematode-feeding microarthropods which frequent the rhizosphere, and fungal-feeding nematodes which apparently interfere with the ability of root-feeding nematodes to find roots. A second result of disturbance is a reduction in the ability of plants to resist nematode feeding. Thus, following a disturbance which disrupts ecosystem stability, root-feeding nematode numbers increase more rapidly than other groups, to the detriment of ecosystem productivity. Thus, as a general indicator of ecosystem health, increases in root feeding nematode numbers suggest serious negative impacts on ecosystem stability.

A decrease in nematode-feeding nematode numbers initiates a trophic cascade effect. For example, nematode-feeding nematodes can control the populations of bacterial-, fungal-or root-feeding nematodes and reduction in nematode-feeders results in an immediate increase in their prey group- i.e., bacterial- fungal- or root feeding nematodes. This in turn results in a decrease in the prey of these three nematode groups; a reduction in bacteria, fungi or roots, each with it's detrimental effect on nutrient cycling Indoor plant growth, as outlined above.

VAM Spore Numbers

Vesicular-arbuscular mycorrhizal (YAM) fungi are critically important for all crop plants, except species of the brassica family (e.g., mustards, kale). A number of researchers have shown that the lack of VAM inoculum, or the lack of the appropriate inoculum can result in poor plant growth, in poor competition with other plants or inability to reproduce or survive under certain extreme conditions. However, most crop fields have adequate VAM spores present, especially if crop residue is placed back into the field. Only in a few situations where soil degradation has been severe, such as with intensive pesticide use, fumigation, or intense fertilizer amendment, will VAM inoculum become so low that plant growth will be in jeopardy.

In restoration studies, the lack of appropriate inoculum is more likely to be a problem than in other situations where sources of appropriate VAM spores are near-by. Thus, the presence of at least 1 to 5 spores per gram of soil is adequate for most crop fields. When the number of spores falls below one per gram, then addition of compost containing high numbers of VAM spores (for example from an alfalfa field, or other legume), or inoculation of VAM spores from a commercial source generally results in positive effects.

Percent VAM Colonization

At least 12% of the root system of grasses, (i.e., most crop plants), should be colonized by VAM in order to obtain the minimum required benefits from this symbiotic relationship. Colonization upwards of 40% is usually seen in healthy soils. VAM colonization can limit root-feeding nematode attack of root systems, if the nematode burden is not too high. A great deal knowledge of the relationship between plant species, VAM species and soil type, including fertility, is needed in order to fully predict the optimal relationship between crop plant, VAM species and soil.

For more information about the Soil Microbial Biomass Service and how to submit a soil sample, write Dr. Elaine Ingham, Department of Botany and Plant Pathology, Cordley Hall 2082, Oregon State University, Corvallis, OR 97331-2902.

Pesticide

A pesticide is a substance or mixture of substances used to kill a pest. A pesticide is any substance or mixture of substance intended for:- preventing, destroying, repelling or mitigating any pest. A pesticide may be a chemical substance, biological agent (such as a virus or bacteria), antimicrobial, disinfectant or device used against any pest. Pests include insects, plant pathogens, weeds, molluscs, birds, mammals, fish, nematodes (roundworms), microbes and people that destroy property, spread or are a vector for disease or cause a nuisance. Although there are benefits to the use of pesticides, there are also drawbacks, such as potential toxicity to humans and other animals. FAO has defined the term of *pesticide* as:

> *any substance or mixture of substances intended for preventing, destroying or controlling any pest, including vectors of human or animal disease, unwanted species of plants or animals causing harm during or otherwise interfering with the production, processing, storage, transport or marketing of food, agricultural commodities, wood and wood products or animal feedstuffs, or substances which may be administered to animals for the control of insects, arachnids or other pests in or on their bodies. The term includes substances intended for use as a plant growth regulator, defoliant, desiccant or agent for thining fruit or preventing the premature fall of fruit, and substances applied to crops either before or after harvest to protect the commodity from deterioration during storage and transport.*

History

Since before 20 BCE, humans have utilized pesticides to protect their crops. The first known pesticide was elemental sulfur dusting used in Sumer about 4,500 years ago. By the 15th century, toxic chemicals such as arsenic, mercury and lead were being applied to crops to kill pests. In the 17th century, nicotine sulfate was extracted from tobacco leaves for use as an insecticide. The 19th century saw the introduction of two more natural pesticides, pyrethrum which is derived from chrysanthemums, and rotenone which is derived from the roots of tropical vegetables. Until the 1950s, arsenic-based pesticides were dominant. , Paul Müller discovered that DDT was a very effective insecticide. Organochlorines such as DDT were dominant, but they were replaced in the US by organophosphates and carbamates by 1975. Since then, pyrethrin compounds have become the dominant insecticide. Herbicides became common in the 1960s, lead by "triazine and other nitrogen-based compounds, carboxylic acids such as 2,4-dichlorophenoxyacetic acid, and glyphosate".

In the 1940s manufacturers began to produce large amounts of synthetic pesticides and their use became widespread. Some sources consider the 1940s and 1950s to have been the start of the "pesticide era." Pesticide use has increased 50-fold since 1950 and 2.3 million tonnes (2.5 million short tons) of industrial pesticides are now used each year. Seventy-five percent of all pesticides in the world are used in developed countries, but use in developing countries is increasing. In 2001 the EPA stopped reporting pesticide use statistics; the only comprehensive study of pesticide use trends was published in 2003 by the National Science Foundation's Centre for Integrated Pest Management.

In the 1960s, it was discovered that DDT was preventing many fish-eating birds from reproducing, which was a serious threat to biodiversity. Rachel Carson wrote the best-selling book *Silent Spring* about biological magnification. The agricultural use of DDT is now banned under the Stockholm Convention on Persistent Organic Pollutants, but it is still used in some developing nations to prevent malaria and other tropical diseases by spraying on interior walls to kill or repel mosquitoes.

Classification

Pesticides can be classified by target organism, chemical structure, and physical state. Pesticides can also be classed as inorganic, synthetic, or biologicals (biopesticides), although the distinction can sometimes blur. Biopesticides include microbial pesticides and biochemical pesticides. Plant-derived pesticides, or "botanicals", have been developing quickly. These include the pyrethroids, rotenoids, nicotinoids, and a fourth group which includes strychnine and scilliroside.

Many pesticides can be grouped into chemical families. Prominent insecticide families include organochlorines, organophosphates, and carbamates. Organochlorine hydrocarbons (e.g. DDT) could be separated into dichlorodiphenylethanes, cyclodiene compounds, and other related compounds. They operate by disrupting the sodium/potassium balance of the never fiber, forcing the nerve to transmit continuously. Their toxicities vary greatly, but they have been phased out because of their persistence and potential to bioaccumulate. Organophosphate and carbamates largely replaced organochlorines. Both operate through inhibiting the enzyme acetylcholinesterase, allowing acetylcholine to transfer nerve impulses indefinitely and causing a variety of symptoms such as weakness or paralysis.

Organophosphates are quite toxic to vertebrates, and have in some cases been replaced by less toxic carbamates. Thiocarbamate and dithiocarbamates are subclasses of carbamates. Prominent families of herbicides include pheoxy and benzoic acid herbicides (e.g. 2,4-D), triazines (e.g. atrazine), ureas (e.g. diuron), and Chloroacetanilides (e.g. alachlor. Phenoxy compounds tend to selectively kill broadleaved weeds rather than grasses.

The phenoxy and benzoic acid herbicides function similar to plant growth hormones, and grow cells without normal cell division, crushing the plants nutrient transport system. Triazines interfere with photsynthesis. Many commonly used pesticides are not included in these families, including glyphosate.

- Algicides or algaecides for the control of algae
- Avicides for the control of birds
- Bactericides for the control of bacteria
- Fungicides for the control of fungi and oomycetes
- Herbicides (e.g. glyphosate) for the control of weeds
- Insecticides (e.g. organochlorines, organophosphates, carbamates, and pyrethroids) for the control of insects-these can be ovicides (substances that kill eggs), larvicides (substances that kill larvae) or adulticides (substances that kill adults)
- Miticides or acaricides for the control of mites
- Molluscicides for the control of slugs and snails
- Nematicides for the control of nematodes
- Rodenticides for the control of rodents
- Virucides for the control of viruses.

Pesticides can be classified based upon their biological mechanism function or application method. Most pesticides work by poisoning pests. A systemic pesticide moves inside a plant following absorption by the plant. With insecticides and most fungicides, this movement is usually upward (through the xylem) and outward. Increased efficiency may be a result. Systemic insecticides which poison pollen and nectar in the flowers may kill needed pollinators such as bees.

In 2009, the development of a new class of fungicides called paldoxins was announced. These work by taking advantage of natural defence chemicals released by plants called phytoalexins, which fungi then detoxify using enzymes. The paldoxins inhibit the fungi's detoxification enzymes. They are believed to be safer and greener.

Uses

Pesticides are used to control organisms which are considered harmful. For example, they are used to kill mosquitoes that can transmit potentially deadly diseases like west nile virus, yellow fever, and malaria. They can also kill bees, wasps or ants that can cause allergic reactions. Insecticides can protect animals from illnesses that can be caused by parasites such as fleas. Pesticides can prevent sickness in humans that could be caused by mouldy food or diseased produce. Herbicides can be used to clear roadside weeds, trees and brush. They can also kill invasive weeds in parks and wilderness areas which may cause environmental damage.

Herbicides are commonly applied in ponds and lakes to control algae and plants such as water grasses that can interfere with activities like swimming and fishing and cause the water to look or smell unpleasant. Uncontrolled pests such as termites and mould can damage structures such as houses. Pesticides are used in grocery stores and food storage facilities to manage rodents and insects that infest food such as grain. Each use of a pesticide carries some associated risk. Proper pesticide use decreases these associated risks to a level deemed acceptable by pesticide regulatory agencies such as the United States Environmental Protection Agency (EPA) and the Pest Management Regulatory Agency (PMRA) of Canada.

Pesticides can save farmers' money by preventing crop losses to insects and other pests; in the US, farmers get an estimated fourfold return on money they spend on pesticides. One study found that not using pesticides reduced crop yields by about 10%. Another study, conducted in 1999, found that a ban on pesticides in the United States may result in a rise of food prices, loss of jobs, and an increase in world hunger.

DDT, sprayed on the walls of houses, is an organochloride that has been used to fight malaria since the 1950s. Recent policy statements by the World Health Organization have given stronger support to this approach. Dr. Arata Kochi, WHO's malaria chief, said, "One of the best tools we have against malaria is indoor residual house spraying. Of the dozen insecticides WHO has approved as safe for house spraying, the most effective is DDT." However, since then, an October 2007 study has linked breast cancer from exposure to DDT prior to puberty. Poisoning may also occur due to use of DDT and other chlorinated hydrocarbons by entering the human food chain when animal tissues are affected. Symptoms include nervous excitement, tremors, convulsions or death. Scientists estimate that DDT and other chemicals in the organophosphate class of pesticides have saved 7 million human lives since 1945 by preventing the transmission of diseases such as malaria, bubonic plague, sleeping sickness, and typhus. However, DDT use is not always effective, as resistance to DDT was identified in Africa as early as 1955, and by 1972 nineteen species of mosquito worldwide were resistant to DDT. A study for the World Health Organization in 2000 from Vietnam established that non-DDT malaria controls were significantly more effective than DDT use. The ecological effect of DDT on organisms is an example of bioaccumulation.

Regulation

In most countries, in order to sell or use a pesticide, it must be approved by a government agency. For example, in the United States, the Environmental Protection Agency (EPA) does so. Complex and costly studies must be conducted to indicate whether the material is safe to use and effective against the intended pest. During the registration process, a label is created which contains directions for the proper use of the material. Based on acute toxicity, pesticides are assigned to a Toxicity Class.

Some pesticides are considered too hazardous for sale to the general public and are designated restricted use pesticides. Only certified applicators, who have passed an exam, may purchase or supervise the application of restricted use pesticides. Records of sales and use are required to be maintained and may be audited by government agencies charged with the enforcement of pesticide regulations.

In Europe, recent EU legislation has been approved banning the use of highly toxic pesticides including those which are carcinogenic, mutagenic or toxic to reproduction, those which are endocrine-disrupting, and those which are persistent, bioaccumulative and toxic (PBT) or very persistent and very bioaccumulative (vPvB). Measures were approved to improve the general safety of pesticides across all EU member states.

Though pesticide regulations differ from country to country, pesticides and products on which they were used are traded across international borders. To deal with inconsistencies in regulations among countries, delegates to a conference of the United Nations Food and Agriculture Organization adopted an International Code of Conduct on the Distribution and Use of Pesticides in 1985 to create voluntary standards of pesticide regulation for different countries. The Code was updated in 1998 and 2002. The FAO claims that the code has raised awareness about pesticide hazards and decreased the number of countries without restrictions on pesticide use.

Two other efforts to improve regulation of international pesticide trade are the United Nations London Guidelines for the Exchange of Information on Chemicals in International Trade and the United Nations Codex Alimentarius Commission. The former seeks to implement procedures for ensuring that prior informed consent exists between countries buying and selling pesticides, while the latter seeks to create uniform standards for maximum levels of pesticide residues among participating countries. Both initiatives operate on a voluntary basis.

Reading and following label directions is required by law in countries such as the US and in limited parts of the rest of the world.

One study found pesticide self-poisoning the method of choice in one third of suicides worldwide, and recommended, among other things, more restrictions on the types of pesticides that are most harmful to humans.

Environmental Effects

Pesticide use raises a number of environmental concerns. Over 98% of sprayed insecticides and 95% of herbicides reach a destination other than their target species, including non-target species, air, water and soil. Pesticide drift occurs when pesticides suspended in the air as particles are carried by wind to other areas, potentially contaminating them. Pesticides are one of the causes of water pollution, and some pesticides are persistent organic pollutants and contribute to soil contamination.

In addition, pesticide use also reduces biodiversity and results in lower soil quality, reduced nitrogen fixation, contribute to pollinator decline, can reduce habitat, especially for birds, and can threaten endangered species.

Health Effects

Pesticides can be dangerous to consumers, workers and close bystanders during manufacture, transport, or during and after use.

The American Medical Association recommends limiting exposure to pesticides and using safer alternatives:

> *Particular uncertainty exists regarding the long-term effects of low-dose pesticide exposures. Current surveillance systems are inadequate to characterize*

potential exposure problems related either to pesticide usage or pesticide-related illnesses...Considering these data gaps, it is prudent...to limit pesticide exposures...and to use the least toxic chemical pesticide or non-chemical alternative.

Farmers and Workers

The World Health Organization and the UN Environment Programme estimate that each year, 3 million workers in agriculture in the developing world experience severe poisoning from pesticides, about 18,000 of whom die. According to one study, as many as 25 million workers in developing countries may suffer mild pesticide poisoning yearly.

There have been many studies of farmers intended to determine health effects of occupational pesticide exposure. Associations between non-Hodgkin lymphoma, leukemia, prostate cancer, multiple myeloma, and soft tissues sarcoma have been reported in studies, with less associations found for other cancers.

Organophosphate pesticides have increased in use, because they are less damaging to the environment and they are less persistent than organochlorine pesticides. These are associated with acute health problems for workers that handle the chemicals, such as abdominal pain, dizziness, headaches, nausea, vomiting, as well as skin and eye problems. Additionally, many studies have indicated that pesticide exposure is associated with long-term health problems such as respiratory problems, memory disorders, dermatologic conditions, cancer, depression, neurological deficits, miscarriages, and birth defects. Summaries of peer-reviewed research have examined the link between pesticide exposure and neurologic outcomes and cancer, perhaps the two most significant things resulting in organophosphate-exposed workers.

According to researchers from the National Institutes of Health (NIH), licensed pesticide applicators who used chlorinated pesticides on more than 100 days in their lifetime were at greater risk of diabetes. One study found that associations between specific pesticides and incident diabetes ranged from a 20 percent to a 200 percent increase in risk. New cases of diabetes were reported by 3.4 percent of those in the lowest pesticide use category compared with 4.6 percent of those in the highest category. Risks were greater when users of specific pesticides were compared with applicators who never applied that chemical.

Consumers

There are concerns that pesticides used to control pests on food crops are dangerous to people who consume those foods. These concerns are one reason for the organic food movement. Many food crops, including fruits and vegetables, contain pesticide residues after being washed or peeled. Chemicals that are no longer used but which are resistant to breakdown for long periods may remain in soil and water and thus in food.

The United Nations Codex Alimentarius Commission has recommended international standards for Maximum Residue Limits (MRLs), for individual pesticides in food.

In the EU, MRLs are set by DG-SANCO. In the US, levels of residues that remain on foods are limited to tolerance levels that are established by the U.S. Environmental Protection Agency—and are considered safe. The EPA sets the tolerances based on the toxicity of the pesticide and its breakdown products, the amount and frequency of pesticide application, and how much of the pesticide (i.e., the residue) remains in or on food by the time it is marketed and prepared. Tolerance levels are obtained using scientific risk assessments that pesticide manufacturers are required to produce by conducting toxicological studies, exposure modelling and residue studies before a particular pesticide can be registered, however, the effects are tested for single pesticides, and there is little information on possible synergistic effects of exposure to multiple pesticide traces in the air, food and water.

A study published by the United States National Research Council in 1993 determined that for infants and children, the major source of exposure to pesticides is through diet. A study in 2006 measured the levels of organophosphorus pesticide exposure in 23 school children before and after replacing their diet with organic food (food grown without synthetic pesticides). In this study it was found that levels of organophosphorus pesticide exposure dropped dramatically and immediately when the children switched to an organic diet.

To reduce the amounts of pesticide residues in food, consumers can wash, peel, and cook their food; trim the fat from meat; and eat a variety of foods to avoid repeat exposure to a pesticide typically used on a given crop. Since organic food use pesticide, purchase of organic food does not mean one can avoid exposure.

Strawberries and tomatoes are the two crops with the most intensive use of soil fumigants. They are particularly vulnerable to several type of diseases, insects, mites, and parasitic worms. In 2003, in California alone, 3.7 million pounds (1,700 metric tons) of metam sodium were used on tomatoes. In recent years other farmers have demonstrated that it is possible to produce strawberries and tomatoes without the use of harmful chemicals and in a cost effective way.

The Public

Exposure routes other than consuming food that contains residues, in particular pesticide drift, are potentially significant to the general public. The Bhopal disaster occurred when a pesticide plant released 40 tons of methyl isocyanate (MIC) gas, a chemical intermediate in the synthesis of some carbamate pesticides. The disaster immediately killed nearly 3,000 people and ultimately caused at least 15,000 deaths.

In China, an estimated half million people are poisoned by pesticides each year, 500 of whom die.

Children have been found to be especially susceptible to the harmful effects of pesticides. A number of research studies have found higher instances of brain cancer, leukemia and birth defects in children with early exposure to pesticides, according to the Natural Resources Defence Council. Often used for ridding school buildings of rodents, insects, pests, etc., pesticides only work temporarily and must be re-applied. The poisons found in pesticides are not selectively harmful to just pests and in everyday school environments children (and faculty) are exposed to high levels of pesticides and cleaning materials. "No testing has ever been done specifically pertaining to threats among children"

Peer-reviewed studies now suggest neurotoxic effects on developing animals from organophosphate pesticides at legally tolerable levels, including fewer nerve cells, lower birth weights, and lower cognitive scores. The United States Environmental Protection Agency—finished a 10 year review of the organophosphate pesticides following the 1996 Food Quality Protection Act, but did little to account for developmental neurotoxic effects, drawing strong criticism from within the agency and from outside researchers.

Some scientists think that exposure to pesticides in the uterus may have negative effects on a fetus that may manifest as problems such as growth and behavioural disorders or reduced resistance to pesticide toxicity later in life.

A new study conducted by the Harvard School of Public Health in Boston, has discovered a 70% increase in the risk of developing Parkinson's disease for people exposed to even low levels of pesticides.

A 2008 study from Duke University found that the Parkinson's patients were 61 percent more likely to report direct pesticide application than were healthy relatives. Both insecticides and herbicides significantly increased the risk of Parkinson's disease.

One study found that use of pesticides may be behind the finding that the rate of birth defects such as missing or very small eyes is twice as high in rural areas as in urban areas. Another study found no connection between eye abnormalities and pesticides. In the USA, increase in birth defects is associated with conceiving in the same period of the year when agrichemicals are in elevated concentrations in surface water.

Pyrethrins, insecticides commonly used in common bug killers, can cause a potentially deadly condition if breathed in.

Continuing Development

Pesticide safety education and pesticide applicator regulation are designed to protect the public from pesticide misuse, but do not eliminate all misuse. Reducing the use of pesticides and choosing less toxic pesticides may reduce risks placed on society and the environment from pesticide use. Integrated pest management, the use of multiple approaches to control pests, is becoming widespread and has been used with success in countries such as Indonesia, China, Bangladesh, the US, Australia, and Mexico. IPM attempts to recognize the more widespread impacts of an action on an ecosystem, so that natural balances are not upset. New pesticides are being developed, including biological and botanical derivatives and alternatives that are thought to reduce health and environmental risks. In addition, applicators are being encouraged to consider alternative controls and adopt methods that reduce the use of chemical pesticides.

Pesticides can be created that are targeted to a specific pest's life cycle, which can be environmentally more friendly. For example, potato cyst nematodes emerge from their protective cysts in response to a chemical excreted by potatoes; they feed on the potatoes and damage the crop. A similar chemical can be applied to fields early, before the potatoes are planted, causing the nematodes to emerge early and starve in the absence of potatoes.

Alternatives

Alternatives to pesticides are available and include methods of cultivation, use of Biological controls, such as pheromones and microbial pesticides, and genetic engineering, and methods of interfering with insect breeding. Application of composted yard waste has also been used as a way of controlling pests. These methods are becoming increasingly popular and often are safer than traditional chemical pesticides. In addition, EPA is registering reduced-risk conventional pesticides in increasing numbers.

Cultivation practices include polyculture (growing multiple types of plants), crop rotation, planting crops in areas where the pests that damage them do not live, timing planting according to when pests will be least problematic, and use of trap crops that attract pests away from the real crop. In the US, farmers have had success controlling insects by spraying with hot water at a cost that is about the same as pesticide spraying.

Release of other organisms that fight the pest is another example of an alternative to pesticide use. These organisms can include natural predators or parasites of the pests. Biological pesticides based on entomopathogenic fungi, bacteria and viruses cause disease in the pest species can also be used.

Interfering with insects' reproduction can be accomplished by sterilizing males of the target species and releasing them, so that they mate with females but do not produce offspring. This technique was first used on the screwworm fly in 1958 and has since been used with the medfly, the tsetse fly, and the gypsy moth. However, this can be a costly, time consuming approach that only works on some types of insects.

Another alternative to pesticides is the thermal treatment of soil through steam. Soil steaming kills pest and increases soil health.

In India, traditional pest control methods include using Panchakavya, the "mixture of five products." The method has recently experienced a resurgence in popularity due in part to use by the organic farming community.

Effectiveness

Some evidence shows that alternatives to pesticides can be equally effective as the use of chemicals. For example, Sweden has halved its use of pesticides with hardly any reduction in crops. In Indonesia, farmers have reduced pesticide use on rice fields by 65% and experienced a 15% crop increase. A study of Maize yields in northern Florida found that the application of composted yard waste with high carbon to nitrogen ratio to agricultural fields was highly effective at reducing the population of plant-parasitic nematodes and increasing crop yield, with yield increases ranging from 10% to 212%; the observed effects were long-term, often not appearing until the third season of the study.

Pesticides as Water Pollutants

The term "pesticide" is a composite term that includes all chemicals that are used to kill or control pests. In agriculture, this includes herbicides (weeds), insecticides (insects), fungicides (fungi), nematocides (nematodes), and rodenticides (vertebrate poisons).

A fundamental contributor to the Green Revolution has been the development and application of pesticides for the control of a wide variety of insectivorous and herbaceous pests that would otherwise diminish the quantity and quality of food produce. The use of pesticides coincides with the "chemical age" which has transformed society since the 1950s. In areas where intensive monoculture is practised, pesticides were used as a standard method for pest control. Unfortunately, with the benefits of chemistry have also come disbenefits, some so serious that they now threaten the long-term survival of major ecosystems by disruption of predator-prey relationships and loss of biodiversity. Also, pesticides can have significant human health consequences.

While agricultural use of chemicals is restricted to a limited number of compounds, agriculture is one of the few activities where chemicals are intentionally released into the environment because they kill things.

Agricultural use of pesticides is a subset of the larger spectrum of industrial chemicals used in modern society. The American Chemical Society database indicates that there were some 13 million chemicals identified in 1993 with some 500 000 new compounds being added annually. In the Great Lakes of North America, for example, the International Joint Commission has estimated that there are more than 200 chemicals of concern in water and sediments of the Great Lakes ecosystem. Because the environmental burden of toxic chemicals includes both agriculture and non-agricultural compounds, it is difficult to separate the ecological and human health effects of pesticides from those of industrial compounds that are intentionally or accidentally released into the environment. However, there is overwhelming evidence that agricultural use of pesticides has a major impact on water quality and leads to serious environmental consequences.

Although the number of pesticides in use is very large, the largest usage tends to be associated with a small number of pesticide products.

In a recent survey in the agricultural western provinces of Canada where some fifty pesticides are in common use, 95% of the total pesticide application is from nine separate herbicides. Although pesticide use is low to nil in traditional and subsistence farming in Africa and Asia, environmental, public health and water quality impacts of inappropriate and excessive use of pesticides are widely documented. For example, Appelgren reports for Lithuania that while pesticide pollution has diminished due to economic factors, water pollution by pesticides is often caused by inadequate storage and distribution of agrochemicals.

In the United States, the US-EPA's National Pesticide Survey found the 10.4% of community wells and 4.2% of rural wells contained detectible levels of one or more pesticides. In a study of groundwater wells in agricultural southwestern Ontario (Canada), 35% of the wells tested positive for pesticides on at least one occasion.

The impact on water quality by pesticides is associated with the following factors:

- Active ingredient in the pesticide formulation.
- Contaminants that exist as impurities in the active ingredient.
- Additives that are mixed with the active ingredient (wetting agents, diluents or solvents, extenders, adhesives, buffers, preservatives and emulsifiers).
- Degradate that is formed during chemical, microbial or photochemical degradation of the active ingredient.

In addition to use of pesticides in agriculture, silviculture also makes extensive use of pesticides. In some countries, such as Canada, where one in ten jobs is in the forest industry, control of forest pests, especially insects, is considered by the industry to be essential. Insecticides are often sprayed by aircraft over very large areas.

Irrigated agriculture, especially in tropical and subtropical environments, usually requires modification of the hydrological regime which, in turn, creates habitat that is conducive to breeding of insects such as mosquitoes which are responsible for a variety of vector-borne diseases. In addition to pesticides used in the normal course of irrigated agriculture, control of vector-borne diseases may require additional application of insecticides such as DDT which have serious and widespread ecological consequences. In order to address this problem, environmental management methods to control breeding of disease vectors are being developed and tested in many irrigation projects.

Historical Development of Pesticides

The history of pesticide development and use is the key to understanding how and why pesticides have been an environmental threat to aquatic systems, and why this threat is diminishing in developed countries and remains a problem in many developing countries. Stephenson and Solomon (1993) outlined the chronology present.

North-south Dilemma over Pesticide Economics

As noted above, the general progression of pesticide development has moved from highly toxic, persistent and bioaccumulating pesticides such as DDT, to pesticides that degrade rapidly in the environment and are less toxic to non-target organisms. The developed countries have banned many of the older pesticides due to potential toxic effects to man and/or their impacts on ecosystems, in favour of more modern pesticide formulations. In the developing countries, some of the older pesticides remain the cheapest to produce and, for some purposes, remain highly effective as, for example, the use of DDT for malaria control. Developing countries maintain that they cannot afford, for reasons of cost and/or efficacy, to ban certain older pesticides. The dilemma of cost/efficacy versus ecological impacts, including long range impacts via atmospheric transport, and access to modern pesticide formulations at low cost remains a contentious global issue.

In addition to ecological impacts in countries of application, pesticides that have been long banned in developed countries (such as DDT, toxaphene, etc.), are consistently found in remote areas such as the high arctic. Chemicals that are applied in tropical and subtropical countries are transported over long distances by global circulation. The global situation has deteriorated to the point where many countries are calling for a global convention on "POPs" (Persistent Organic Pollutants) which are mainly chlorinated compounds that exhibit high levels of toxicity, are persistent, and bioaccumulate. The list is not yet fixed; however, "candidate" substances include several pesticides that are used extensively in developing countries.

Fate and Effects of Pesticides

Factors affecting pesticide toxicity in aquatic systems.

The ecological impacts of pesticides in water are determined by the following criteria:

- *Toxicity:* Mammalian and non-mammalian toxicity usually expressed as LD_{50} ("Lethal Dose": concentration of the pesticide which will kill half the test organisms over a specified test period). The lower the LD_{50}, the greater the toxicity; values of 0-10 are extremely toxic (OMAF, 1991).

Drinking water and food guidelines are determined using a risk-based assessment. Generally, Risk = Exposure (amount and/or duration) × Toxicity.

Toxic response (effect) can be acute (death) or chronic (an effect that does not cause death over the test period but which causes observable effects in the test organism such as cancers and tumours, reproductive failure, growth inhibition, teratogenic effects, etc.).

- *Persistence:* Measured as half-life (time required for the ambient concentration to decrease by 50%). Persistence is determined by biotic and abiotic degradational processes. Biotic processes are biodegradation and metabolism; abiotic processes are mainly hydrolysis, photolysis, and oxidation (Calamari and Barg, 1993). Modern pesticides tend to have short half lives that reflect the period over which the pest needs to be controlled.

- *Degradates:* The degradational process may lead to formation of "degradates" which may have greater, equal or lesser toxicity than the parent compound. As an example, DDT degrades to DDD and DDE.

- *Fate (Environmental):* The environmental fate (behaviour) of a pesticide is affected by the natural affinity of the chemical for one of four environmental compartments (Calamari and Barg, 1993): solid matter (mineral matter and particulate organic carbon), liquid (solubility in surface and soil water), gaseous form (volatilization), and biota. This behaviour is often referred to as "partitioning" and involves, respectively, the determination of: the soil sorption coefficient (K_{OC}); solubility; Henry's Constant (H); and the n-octanol/water partition coefficient (K_{OW}). These parameters are well known for pesticides and are used to predict the environmental fate of the pesticide.

An additional factor can be the presence of impurities in the pesticide formulation but that are not part of the active ingredient. A recent example is the case of TFM, a lampricide used in tributaries of the Great Lakes for many years for the control of the sea lamprey.

Although the environmental fate of TFM has been well known for many years, recent research by Munkittrick *et al.* (1994) has found that TFM formulation includes one or more highly potent impurities that impact on the hormonal system of fish and cause liver disease.

Human Health Effects of Pesticides

Perhaps the largest regional example of pesticide contamination and human health is that of the Aral Sea region. UNEP (1993) linked the effects of pesticides to "the level of oncological (cancer), pulmonary and haematological morbidity, as well as on inborn deformities... and immune system deficiencies".

Human health effects are caused by:

* Skin contact: handling of pesticide products
* Inhalation: breathing of dust or spray
* Ingestion: pesticides consumed as a contaminant on/in food or in water.

Farm workers have special risks associated with inhalation and skin contact during preparation and application of pesticides to crops. However, for the majority of the population, a principal vector is through ingestion of food that is contaminated by pesticides. Degradation of water quality by pesticide runoff has two principal human health impacts. The first is the consumption of fish and shellfish that are contaminated by pesticides; this can be a particular problem for subsistence fish economies that lie downstream of major agricultural areas.

The second is the direct consumption of pesticide-contaminated water. WHO (1993) has established drinking water guidelines for 33 pesticides. Many health and environmental protection agencies have established "acceptable daily intake" (ADI) values which indicate the maximum allowable daily ingestion over a person's lifetime without appreciable risk to the individual. For example, in a recent paper by Wang and Lin (1995) studying substituted phenols, tetrachlorohydroquinone, a toxic metabolite of the biocide pentachlorophenol, was found to produce "significant and dose-dependent DNA damage".

Ecological Effects of Pesticides

Pesticides are included in a broad range of organic micro pollutants that have ecological impacts. Different categories of pesticides have different types of effects on living organisms, therefore generalization is difficult. Although terrestrial impacts by pesticides do occur, the principal pathway that causes ecological impacts is that of water contaminated by pesticide runoff. The two principal mechanisms are bioconcentration and biomagnification.

Bioconcentration: This is the movement of a chemical from the surrounding medium into an organism. The primary "sink" for some pesticides is fatty tissue ("lipids"). Some pesticides, such as DDT, are "lipophilic", meaning that they are soluble in, and accumulate in, fatty tissue such as edible fish tissue and human fatty tissue. Other pesticides such as glyphosate are metabolized and excreted.

Biomagnification: This term describes the increasing concentration of a chemical as food energy is transformed within the food chain. As smaller organisms are eaten by larger organisms, the concentration of pesticides and other chemicals are increasingly magnified in tissue and other organs. Very high concentrations can be observed in top predators, including man.

The ecological effects of pesticides (and other organic contaminants) are varied and are often interrelated. Effects at the organism or ecological level are usually considered to be an early warning indicator of potential human health impacts. The major types of effects are listed below and will vary depending on the organism under investigation and the type of pesticide. Different pesticides have markedly different effects on aquatic life which makes generalization very difficult. The important point is that many of these effects are chronic (not lethal), are often not noticed by casual observers, yet have consequences for the entire food chain.

- Death of the organism.
- Cancers, tumours and lesions on fish and animals.
- Reproductive inhibition or failure.
- Suppression of immune system.
- Disruption of endocrine (hormonal) system.
- Cellular and DNA damage.
- Teratogenic effects (physical deformities such as hooked beaks on birds).
- Poor fish health marked by low red to white blood cell ratio, excessive slime on fish scales and gills, etc.

- Intergenerational effects (effects are not apparent until subsequent generations of the organism).
- Other physiological effects such as egg shell thinning.

These effects are not necessarily caused solely by exposure to pesticides or other organic contaminants, but may be associated with a combination of environmental stresses such as eutrophication and pathogens. These associated stresses need not be large to have a synergistic effect with organic micro pollutants.

Ecological effects of pesticides extend beyond individual organisms and can extend to ecosystems. Swedish work indicates that application of pesticides is thought to be one of the most significant factors affecting biodiversity. Jonsson *et al.* (1990) report that the continued decline of the Swedish partridge population is linked to changes in land use and the use of chemical weed control.

Chemical weed control has the effect of reducing habitat, decreasing the number of weed species, and of shifting the balance of species in the plant community. Swedish studies also show the impact of pesticides on soil fertility, including inhibition of nitrification with concomitant reduced uptake of nitrogen by plants (Torstensson, 1990).

These studies also suggest that pesticides adversely affect soil microorganisms which are responsible for microbial degradation of plant matter (and of some pesticides), and for soil structure. Some regional examples of ecological effects of pesticides.

Natural Factors that Degrade Pesticides

In addition to chemical and photochemical reactions, there are two principal biological mechanisms that cause degradation of pesticides. These are (1) microbiological processes in soils and water and (2) metabolism of pesticides that are ingested by organisms as part of their food supply. While both processes are beneficial in the sense that pesticide toxicity is reduced, metabolic processes do cause adverse effects in, for example, fish.

Energy used to metabolize pesticides and other xenobiotics (foreign chemicals) is not available for other body functions and can seriously impair growth and reproduction of the organism.

Degradation of Pesticides in Soil: "Many pesticides dissipate rapidly in soils. This process is mineralization and results in the conversion of the pesticide into simpler compounds such H_2O, CO_2, and NH_3. While some of this process is a result of chemical reactions such as hydrolysis and photolysis, microbiological catabolism and metabolism is usually the major route of mineralization.

Soil micro biota utilize the pesticide as a source of carbon or other nutrients. Some chemicals (for example 2,4-D) are quite rapidly broken down in soil while others are less easily attacked (2,4,5-T). Some chemicals are very persistent and are only slowly broken down (atrazine)".

Process of Metabolism: Metabolism of pesticides in animals is an important mechanism by which organisms protect themselves from the toxic effects of xenobiotics (foreign chemicals) in their food supply. In the organism, the chemical is transformed into a less toxic form and either excreted or stored in the organism. Different organs, especially the liver, may be involved, depending on the chemical. Enzymes play an important role in the metabolic process and the presence of certain enzymes, especially "mixed" function oxygenases (MFOs) in liver, is now used as an indicator that the organism has been exposed to foreign chemicals.

Pesticide Monitoring in Surface Water

Monitoring data for pesticides are generally poor in much of the world and especially in developing countries. Key pesticides are included in the monitoring schedule of most western countries, however the cost of analysis and the necessity to sample at critical times of the year often preclude development of an extensive data set. Many developing countries have difficulty carrying out organic chemical analysis due to problems of inadequate facilities, impure reagents, and financial constraints. New techniques using immunoassay procedures for presence/absence of specific pesticides may reduce costs and increase reliability. Immunoassay tests are available for triazines, acid amides, carbamates, 2,4-D/phenoxy acid, paraquot and aldrin.

Data on pesticide residues in fish for lipophilic compounds, and determination of exposure and/or impact of fish to lipophobic pesticides through liver and/or bile analysis is mainly restricted to research programmes. Hence, it is often difficult to determine the presence, pathways and fate of the range of pesticides that are now used in large parts of the world.

In contrast, the ecosystemic impacts from older, organochlorine pesticides such as DDT, became readily apparent and has resulted in the banning of these compounds in many parts of the world for agricultural purposes.

Table 1: Proportion of selected pesticides found in association with suspended sediment (After Ongley *et al.*, 1992)

Pesticide	log K_{OW}	% of chemical load at different concentrations (mg/l) of suspended sediment			
		mg/l = 10	mg/l = 100	mg/l = 1000	mg/l = 10000
Aldrin	5.5	15	55	90	100
Atrazine	2.6	0	0	2	20
Chlordane	6.0	30	75	95	100
DDT	5.8	20	67	93	100
Dieldrin	5.5	15	55	90	100
Endrin	5.6	18	57	90	100
Endosulfan	3.6	0	0	21	57
Heptachlor	5.4	13	48	88	100
Lindane	3.9	0	2	30	80
Mirex	6.9	75	95	100	100
Toxaphene[1]	3.3	0	0	12	47
Trifluralin	5.3	12	45	87	100
2,4-D	2.0[2]	0	0	0	4

[1] Toxaphene mixture.

[2] Range is 1.5-2.5.

Table below indicates why older pesticides, together with other hydrophobic carcinogens such as PAHs and PCBs, are poorly monitored when using water samples. As an example, the range of concentration of suspended solids in rivers is often between 100 and 1000 mg/l except during major runoff events when concentrations can greatly exceed these values. Tropical rivers that are unimpacted by development have very low suspended sediment concentrations, but increasingly these are a rarity due to agricultural expansion and deforestation in tropical countries.

As an example, approximately 67% of DDT is transported in association with suspended matter at sediment concentrations as low as 100 mg/l, and increases to 93% at 1000 mg/l of suspended sediment. Given the analytical problems of inadequate detection levels and poor quality control in many laboratories of the developing countries, plus the fact that recovery rates (part of the analytical procedure) can vary from 50-150% for organic compounds, it follows that monitoring data from water samples are usually a poor indication of the level of pesticide pollution for compounds that are primarily associated with the solid phase. The number of NDs (Not Detectable) in many databases is almost certainly an artifact of the wrong sampling medium (water) and, in some cases, inadequate analytical facilities and procedures. Clearly, this makes pesticide assessment in water difficult in large parts of the world. Experience suggests that sediment-associated pesticide levels are often much higher than recorded, and NDs are often quite misleading. Some water quality agencies now use multimedia (water + sediment + biota) sampling in order to more accurately characterize pesticides in the aquatic environment.

Another problem is that analytical detection levels in routine monitoring for certain pesticides may be too high to determine presence/absence for protection of human health. Gilliom (1984) noted that the US Geological Survey's Pesticide Monitoring Network [in 1984] had a detection limit of 0.05 m g/l for DDT, yet the aquatic life criterion is 0.001 m g/l and the human health criterion is 0.0002 m g/l- both much less that the routine detection limit of the programme.

ND (not detectible) values, therefore, are not evidence that the chemical is not present in concentrations that may be injurious to aquatic life and to human health. That this analytical problem existed in the United States suggests that the problem of producing water quality data that can be used for human health protection from pesticides in developing countries, must be extremely serious. Additionally, detection limits are only one of many analytical problems faced by environmental chemists when analysing for organic contaminants.

Even when one has good analytical values from surface water and/or sediments, the interpretation of pesticide data is not straight forward. For example, the persistence of organochlorine pesticides is such that the detection of, say, DDT may well indicate only that (1) the chemical has been deposited through long range transport from some other part of the world, or (2) it is a residual from the days when it was applied in that region. In North America, for example, DDT is still routinely measured even though it has not been used for almost two decades. The association of organochlorine pesticides with sediment means that the ability of a river basin to cleanse itself of these chemicals is partly a function of the length of time it requires for fine-grained sediment to be transported through the basin. Geomorphologists now know that the process of erosion and transport of silts and clays is greatly complicated by sedimentation within the river system and that this fine-grained material may take decades to be transported out of the river basin. For sediment-associated and persistent pesticides that are still in use in some countries, the presence of the compound in water and/or sediments results from a combination of current and past use. As such, the data make it difficult to determine the efficacy of policy decisions such as restrictive use or bans.

Pesticide monitoring requires highly flexible field and laboratory programmes that can respond to periods of pesticide application, which can sample the most appropriate medium (water, sediment, biota), are able to apply detection levels that have meaning for human health and ecosystem protection, and which can discriminate between those pesticides which appear as artifacts of historical use versus those that are in current use.

For pesticides that are highly soluble in water, monitoring must be closely linked to periods of pesticide use. In the United States where there have been major studies of the behaviour of pesticide runoff, the triazines (atrazine and cyanazine) and alachlor (chlorinated acetamide) are amongst the most widely used herbicides. These are used mainly in the spring (May). Studies by Schottler *et al.* (1994) indicate that 55-80% of the pesticide runoff occurred in the month of June.

The significance for monitoring is that many newer and soluble pesticides can only be detected shortly after application; therefore, monitoring programmes that are operated on a monthly or quarterly basis (typical of many countries) are unlikely to be able to quantify the presence or determine the significance of pesticides in surface waters. Pesticides that have limited application are even less likely to be detected in surface waters. The danger lies in the presumption by authorities that ND (non-detectable) values implies that pesticides are absent. It may well only mean that monitoring programmes failed to collect data at the appropriate times or analysed the wrong media.

Pesticide Management and Control

Prediction of water quality impacts of pesticides and related land management practices is an essential element of site-specific control options and for the development of generic approaches for pesticide control. Prediction tools are mainly in the form of models, many of which are containe. Also, the key hydrological processes that control infiltration and runoff, and erosion and sediment transport, are controlling factors in the movement of pesticides.

The European Experience

The Netherlands National Institute of Public Health and Environmental Protection (RIVM, 1992) concluded that "groundwater is threatened by pesticides in all European states. This is obvious both from the available monitoring data and calculations concerning pesticide load, soil sensitivity and leaching... It has been calculated that on 65% of all agricultural land the EC standard for the sum of pesticides (0.5; m g/l) will be exceeded. In approximately 25% of the area this standard will be exceeded by more than 10 times..."

In recognition of pesticide abuse and of environmental and public health impacts the European countries have adopted a variety of measures that include the following (FAO/ECE, 1991):

- Reduction in use of pesticides (by up to 50% in some countries).
- Bans on certain active ingredients.
- Revised pesticide registration criteria.
- Training and licensing of individuals that apply pesticides.
- Reduction of dose and improved scheduling of pesticide application to more effectively meet crop needs and to reduce preventative spraying.
- Testing and approval of spraying apparatus.
- Limitations on aerial spraying.
- Environmental tax on pesticides.
- Promote the use of mechanical and biological alternatives to pesticides.

Elsewhere, as for example Indonesia, reduction in subsidies has reduced the usage of pesticides and has increased the success of integrated pesticide management programmes.

Pesticide Registration

Pesticide control is mainly carried out by a system of national registration which limits the manufacture and/or sale of pesticide products to those that have been approved. In developed countries, registration is a formal process whereby pesticides are examined, in particular, for mammalian toxicity (cancers, teratogenic and mutagenic effects, etc.) and for a range of potential environmental effects based on the measured or estimated environmental behaviour of the product based on its physico-chemical properties. Most developing countries have limited capability to carry out their own tests on pesticides and tend to adopt regulatory criteria from the developed world. As our knowledge of the effects of pesticides in the environment accumulates, it has become apparent that many of the older pesticides have inadequate registration criteria and are being re-evaluated. As a consequence, the environmental effects of many of the older pesticides are now recognized as so serious that they are banned from production or sale in many countries.

A dilemma in many developing countries is that many older pesticides (e.g. DDT) are cheap and effective. Moreover, regulations are often not enforced with the result that many pesticides that are, in fact, banned, are openly sold and used in agricultural practice. The dichotomy between actual pesticide use and official policy on pesticide use is, in many countries, far apart.

Regulatory control in many countries is ineffective without a variety of other measures, such as education, incentives, etc. The extent to which these are effective in developed versus developing countries depends very much on (1) the ability of government to effectively regulate and levy taxes and (2) on the ability or readiness of the farming community to understand and act upon educational programmes.

The fundamental dilemma remains one of accommodating local and short term gain by the farmer (and manufacturer and/or importer) by application of an environmentally dangerous pesticides, with societal good by the act of limiting or banning its use.

There is now such concern over environmental and, in some instances, human health effects of excessive use and abuse of pesticides, that there is active discussion within many governments of the need to include a programme of pesticide reduction as part of a larger strategy of sustainable agriculture. In 1992, Denmark, the Netherlands and Sweden were the first of the 24 member states of the OECD to embark upon such a programme.

The Netherlands is the world's second largest exporter of agricultural produce after the United States. In contrast, wood preservatives in the forest sector account for 70% of Swedish pesticide use with agriculture using only 30%. As noted above, the lack of baseline data on pesticides in surface waters of OECD countries, is a constraint in establishing baseline values against which performance of the pesticide reduction programme can be measured.

Information on EXTOXNET, a pesticide-toxicology network which is available on INTERNET.

The Danish example:

In 1986 the Danish Government initiated an Action Plan for sustainable agriculture which would prevent the use of pesticides for two purposes :

- Safeguard human health-from the risks and adverse effects associated with the use of pesticides, primarily by preventing intake via food and drinking water.

- Protect the environment-both the non-target and beneficial organisms found in the flora and fauna on cultivated land and in aquatic environments.

The objective was to achieve a 50% reduction in the use of agricultural pesticides by 1997 from the average amount of pesticides used during the period 1981-85. This was to be measured by (1) a decline in total sales (by weight) of the active ingredients and, (2) decrease in frequency of application.

While the World Wide Fund for Nature report that by 1993, sales of active ingredients had been reduced by 30%, the application frequency had not declined.

The Danish legislation included the following components although, by 1993, not all had achieved comparable success.

- *Reassessment of active ingredients:* Reassessment reflects improved scientific knowledge of pathways, fate and effects of pesticides. By 1993, 80% of the 223 active ingredients had been reassessed. Fewer than 40% had been approved and about 15% are restricted to specific types of application (WWF summary of Danish Environmental Protection reports).

- *Promotion of organic agriculture:* The legislation included funding to promote conversion of traditional agriculture to organic agriculture which, by definition, does not use pesticides.

- *Excise tax on pesticides:* The Danish Institute of Agriculture concluded that, "A tax on pesticides can be designed and implemented in such a way that it will reduce the use of pesticides without distorting or dramatically worsening the economic situation in the agricultural sector." Funds raised by the tax were to be directed back to the agricultural sector. Studies reported by the Institute of Agriculture suggested, however, that pesticide taxes alone would not produce the requisite reduction during the lifetime of the plan.

- *Certification of pesticide users:* All farmers and commercial sprayers must hold application certificates. Certification includes education in pesticide issues.

- Records of pesticide application: Commencing 1 August 1993, individual farmers were required to maintain records of pesticide application.

- *Approval of spraying equipment:* This measure gives the Ministry of Agriculture some control of types of spraying equipment used in Denmark. New computer controlled sprayers permit continuous monitoring of pesticide dose by the farmer and reduces excessive application.

The Danish Government is considering the following additional components as part of the regulatory process:

- *Maximum limits on the environmental load of pesticides:* The intent is to produce an index which equates the quantity used of a pesticide with its known ecological effects. The concept is, however, difficult to implement as noted by the WWF, "... there is no direct relationship between the pesticide-load index and the environmental effects-direct or indirect-of pesticides, since these are the result of a complex interaction between many different factors." Nevertheless, the concept has certain management and regulatory value and may be possible, initially, with a few common pesticides.

- *Prohibiting the use of pesticides within 10 m of lakes, watercourses, wetlands, and conservation areas:* This would achieve some level of pesticide protection for aquatic systems in the same manner that buffer strips are widely used to reduce the effects of sedimentation.

- *Prohibiting the use of pesticides within a specified distance from private gardens and properties containing fields that are cultivated without the use of pesticides.*

- *Prohibiting the use of pesticides within 10 m of a drinking-water reservoir.*

The Swedes have had considerable success in achieving pesticide reduction targets. WWF (1992) credits their success to the following factors.

- Setting of targets with achievable goals and using multiple measures of reduction.

- Lead role played by the Environment Ministry and Chemicals Inspectorate.
- Active support of farmers organizations which realize the economic and environmental advantages of reduced pesticide usage.
- A strong research and development base that provides credible support for new pesticide initiatives.
- Certification of new machinery and routine testing of farm sprayers at government-regulated test centres.
- Re-evaluation and re-registration of pesticides which has resulted in 338 products being removed from the market.

Chapter 4

Soil Analysis Terms and Testing

Soil pH

The soil pH measures active soil acidity or alkalinity. A pH of 7.0 is neutral. Values lower than 7.0 are acid; values higher are alkaline. Usually the most desirable pH range for mineral soils is 6.0 to 7.0 and for organic soils 5.0 to 5.5. The soil pH is the value that should be maintained in the pH range most desirable for the crop to be grown.

Buffer pH

This is an index value used for determining the amount of lime to apply on acid soils to bring the pH to the desired pH for the crop to be grown. The lower the buffer pH reading the higher the lime requirement.

Phosphorus

The phosphorus test measures that phosphorus that should be available to the plant. The optimum level will vary with crop, yield and soil conditions, but for most field crops a medium to optimum rating is adequate. For soils with pH above 7.3 the sodium bicarbonate test will determine the available P.

Potassium

This test measures available potassium. The optimum level will vary with crop, yield, soil type, soil physical condition, and other soil related factors. Generally higher levels of potassium are needed on soils high in clay and organic matter versus soils, which are sandy and low in organic matter. Optimum levels for light-coloured, coarse-textured soils may range from 90 to 125 ppm (180 to 250 lbs/ac). On dark-coloured heavy-textured soils levels ranging from 125 to 200 ppm (250 to 400 lbs/ac) may be required.

Calcium

Primarily soil type, drainage, liming and cropping practices affect the levels of calcium found in the soil.

Calcium is closely related to soil pH. Calcium deficiencies are rare when soil pH is adequate. The level for calcium will vary with soil type, but optimum ranges are normally in the 65% to 75% cation saturation range.

Magnesium

The same factors, which affect calcium levels in the soil, also influence magnesium levels except magnesium deficiencies are more common. Adequate magnesium levels range from 30 to 70 ppm (60 to 140 lbs/ac). The cation saturation for magnesium should be 10 to 15%.

Sulphur

The soil test measures sulfate-sulfur. This is a readily available form preferred by most plants. Soil test levels should be maintained in the optimum range.

It's important that other soil factors, including organic matter content, soil texture and drainage be taken into consideration when interpreting sulfur soil test and predicting crop response.

Boron

The readily soluble boron is extracted from the soil. Boron will most likely be deficient in sandy soils, low in organic matter with adequate rainfall. Soil pH, organic matter level and texture should be considered in interpreting the boron test, as well as the crop to be grown.

Copper

Copper is most likely to be deficient on low organic matter sandy soils, or organic soils. The crop to be grown, soil texture, and organic matter should be considered when interpreting copper tests. A rating of medium to optimum should be maintained.

Iron

Soil pH is a very important factor in interpreting iron tests. In addition, crops vary a great deal in sensitivity to iron deficiency. Normally a medium level would be adequate for most soils. If iron is needed it would be best applied foliar.

Manganese Soil

pH is especially important in interpreting manganese test levels. In addition, soil organic matter, crop and yield levels must be considered. Manganese will work best if applied foliar or banded in the soil.

Zinc

Other factors, which should be considered in interpreting the zinc test, include available phosphorus, pH, and crop and yield level. For crops that have a good response to zinc, the soil test level should be optimum.

Sodium

Sodium is not an essential plant nutrient but is usually considered in light of its effect on the physical condition of the soil. Soils high in exchangeable sodium may cause adverse physical and chemical conditions to develop in the soil. These conditions may prevent the growth of plants. Reclamation of these soils involves the replacement of the exchangeable sodium by calcium and the removal by leaching.

Soluble Salts

Excessive concentration of various salts may develop in soils. This may be a natural occurrence or it may result from irrigation, excessive fertilization or contamination from various chemicals or industrial wastes. One effect of high soil salt concentration is to produce water stress in a crop to where plants may wilt or even die. The effect of salinity is negligible if the reading is less than 1.0 mmhos/cm. Readings greater than 1.0 mmhos/cm may affect salt sensitive plants and readings greater than 2.0 mmhos/cm may require the planting of salt tolerant plants.

Organic Matter and ENR (Estimated Nitrogen Release)

Percent organic matter is a measurement of the amount of plant and animal residue in the soil. The colour of the soil is usually closely related to its organic matter content, with darker soils being higher in organic matter. The organic matter serves as a reserve for many essential nutrients, especially nitrogen. During the growing season, a part of this reserve nitrogen is made available to the plant through bacterial activity. The ENR is an estimate of the amount of nitrogen (lbs/acre) that will be released over the season. In addition to organic matter level, this figure may be influenced by seasonal variation in weather conditions as well as soil physical conditions.

NO_3-N (Nitrate Nitrogen)

Nitrate nitrogen is a measure of the nitrogen available to the plant in nitrate form. In high rainfall areas, sandy soil types and areas with warm winters, this measurement may be of limited value except at planting or side dress time. In the areas with lower rainfall, the nitrate test may be very beneficial.

Cation Exchange Capacity (CEC)

Cation exchange capacity measures the soil's ability to hold nutrients such as calcium, magnesium, and potassium, as well as other positively charged ions such as sodium and hydrogen. The CEC of a soil is dependent upon the amounts and types of clay minerals and organic matter present. The common expression for CEC is in terms of milliequivalents per 100 grams (meq/100g) of soil.

The CEC of soil can range from less than 5 to 35 meq/100g for agricultural type soils. Soils with high CEC will generally have higher levels of clay and organic matter. For example, one would expect soil with a silty clay loam texture to have a considerably higher CEC than a sandy loam soil. Although high CEC soils can hold more nutrients, it doesn't necessarily mean that they are more productive. Much depends on good soil management.

Cation Saturation

Cation saturation refers to the proportion of the CEC occupied by a given cation (an ion with a positive charge such as calcium, magnesium or potassium). The percentage saturation for each of the cations will usually be within the following ranges:

Calcium: 40 to 80 percent

Magnesium: 10 to 40 percent

Potassium: 1 to 5 percent

Soil Testing

The farmers find it extremely difficult to know the proper type of fertilizer, which would match his soil. In using a fertilizer he must take into account the requirement of his crops and the characteristics of the soil.

The basic objective of the soil-testing programme is to give farmers a service leading to better and more economic use of fertilizers and better soil management practices for increasing agricultural production. High crop yields cannot be obtained without applying sufficient fertilizers to overcome existing deficiencies.

Efficient use of fertilizers is a major factor in any programme designed to bring about an economic increase in agricultural production. The farmers involved in such a programme will have to use increasing quantities of fertilizers to achieve the desired yield levels. However the amounts and kinds of fertilizers required for the same crop vary from soil to soil, even field to field on the same soil.

The use of fertilizers without first testing the soil is like taking medicine without first consulting a physician to find out what is needed. It is observed that the fertilizers increase yields and the farmers are aware of this. But are they applying right quantities of the right kind of fertilizers at the right time at the right place to ensure maximum profit?

Without a fertilizer recommendation based upon a soil test, a farmer may be applying too much of a little needed plant food element and too little of another element which is actually the principal factor limiting plant growth. This not only means an uneconomical use of fertilizers, but in some cases crop yields actually may be reduced because of use of the wrong kinds or amounts, or improper use of fertilizers. A fertilizers recommendation from a soil testing laboratory is based on carefully conducted soil analyses and the results of up-to-date agronomic research on the crop, and it therefore is most scientific information available for fertilizing that crop in that field.

Each recommendation based on a soil test takes into account the values obtained by these accurate analysis, the research work so far conducted on the crop in the particular soil areas, and the management practices of the concerned farmer.

The soil test with the resulting fertilizer recommendation is therefore the actual connecting link between agronomic research and its practical application to the farmers' fields. However, soil testing is not an end in itself. It is a means to an end. A farmer who follows only the soil test recommendations is not assured of a good crop. Good crop yields are the result of the application also of other good management practices, such as proper tillage, efficient water management, good seed, and adequate plant protection measures. Soil testing is essential and is the first step in obtaining high yields and maximum returns from the money invested in fertilizers.

How to Collect a Soil Sample

1. Sample each field separately. However, where the areas within a field differ distinctly in crop growth, appearance of the soils,

or in elevation, or are known to have been cropped or fertilized and manured differently, divided the filed and sample each area separately.

2. Take a composite sample from each area. Scrape away surface litter, then take a small sample from the surface to plough depth from a number of spots in the field (10 to 15 per acre). Collect these samples in a clean bucket or some such wide container.

3. Where crops have been planted in lines (rows), sample between the lines.

4. Do not sample unusual area. Avoid areas recently fertilized, old bunds, marshy spots, near tress, compost piles, other non-representative locations.

5. Take a uniform thick sample from the surface to plough depth. If a spade or a trowel is used, dig a v-shaped hole, then cut out a uniform thick slice of soil from bottom to op of the exposed soil face, collect the sample on the baled or in your hand and place it in the bucket.

6. Pour the soil from the bucket on a piece of clean cloth or paper and mix thoroughly, discard, by quartering, all but 1 to 2 lbs. of soil. Quarterly may be done by mixing sample well, dividing it into four equal parts, then rejecting two opposite quarters, mixing the remaining two portions, again dividing into four parts and rejecting two opposite quarters, and so on. The sample should be dried in the shade for an hour or two before it goes into the cloth bag container.

7. Each cloth bag should be large enough to hold a pound or two of soil, and should be properly marked to identify the sample.

8. Fill out the soil sample information sheet for each sample. These forms may be sent separately to the laboratory or enclosed with the soil sample.

9. Address the samples to the Soil Chemist, Soil Testing Laboratory, Goal Ghar, Port Blair.

10. Keep a record of the areas sampled and a simple sketch map for reference when you get the soil test and fertilizers recommendation report from the soil testing laboratory.

Role of the Extension Service in Soil Testing

The actual analysis of the sample and the making out of fertilizer recommendation is only part of the soil testing service. To a large measure, the efficiency of this service depends upon the care and effort put froth by extension workers and farmers in the collection and dispatch of samples to the laboratory. Its effectiveness also depends upon the proper follow-through of the fertilizer recommendations, including the establishment of result demonstrations on farmer's fields to induce the farmers to follow the fertilizer recommendations. In this work the staff of the extension service play the most important role, since they are the people directly in contact with the farmers or this reason, the soil chemist in charge of the laboratory must give periodic and through training to the extension staff on these subjects.

Collection of Samples

A useful soil testing service starts with the collection of representative soil samples. A fertilizer recommendation made after analyzing the soil can only as good as the sample on which it is based. Actually the one to ten grams of soil used for each chemical analysis should represent as accurately as possible the entire surface six inches of soil, weighing about 2 million pounds per acre. The importance of taking a representative composite sample is, therefore, self-evident. One field can be treated as a single sampling unit only if it is relatively uniform and does not exceed approximately five acres. Variations in slope, colour, texture, management, and cropping pattern should be taken into account and separate composite soil sample adequately representing the field, small portions of surface soil should be collected to depth of six inches from at least ten well-distributed spots in the field, mixed well, and about ½ kg of representative sample sent to laboratory.

Proper sampling tools are essential for collection of good soil samples. For a soft. Moist soil, the soil tube, phowda (spade), or khurpi (trowel) are usually quite satisfactory.

For harder soils, a screw type auger, or an adze might be more convenient. Post hole augers are convenient for sampling excessively wet areas like paddy fields. An extension worker whose duties include collection of soil samples should be supplied with at least a few of these tools, and also a plastic bucket. The phowda, khurpi and adze are very common implements available in most hardware shops and so there should be no difficulty in procuring these implements.

The farmers should be given help in filling out the soil sample information sheet with an ex-plantation of any items not understood. It should be remembered that the information sheet is very vital part of procedures that go to make a good soil test recommendation. This sheet must supply all of the background information that, in combination with the results of the analysis, makes possible an accurate fertilizer recommendation for a certain crop, for that particular field.

Factors such as crop variety, slope of land, irrigation and drainage facilities, and pervious cropping seasons affect the amounts of fertilizer to be applied to particular crop.

Any peculiarities noted in the soil or in the vigor or the crop would be very valuable information on the soil sample information sheet as a basis for making an adequate fertilizer recommendation. In the absence of this information, the soil chemist must base his recommendation upon the soil test values alone, and more often than not the farmer will receive an adequate fertilizer recommendation.

Soil Analysis: A key to Soil Nutrient Mangement

High yields of top-quality crops require an abundant supply of 16 essential nutrient elements. In addition to providing a place for crops to grow, soil is the source for most of the essential nutrients required by the crop. Our soil resource can be compared to a bank where continued withdrawal without repayment cannot continue indefinitely. As nutrients are removed by one crop and not replaced for subsequent crop production, yields will decrease accordingly. Accurate accounting of nutrient removal and replacement, crop production statistics, and soil analysis results will help the producer manage fertilizer applications.

A soil analysis is used to determine the level of nutrients found in a soil sample. As such, it can only be as accurate as the sample taken in a particular field. The results of a soil analysis provide the agricultural producer with an estimate of the amount of fertilizer nutrients needed to supplement those in the soil. Applying the appropriate type and amount of needed fertilizer will give the agricultural a more reasonable chance to obtain the desired crop yield.

Objectives of Soil Analysis

- To provide an index of nutrient availability or supply in a given soil. The soil extract is designed to evaluate a portion of the nutrients from the same "pool" used by the plant.

- To predict the probability of obtaining a profitable response to fertilizer application. Low analysis soils may not always respond to fertilizer applications due to other limiting factors. However, the probability of a response is greater than on a high analysis soil.
- To provide a basis for fertilizer recommendations for a given crop.
- To evaluate the fertility status of the soil and plan a nutrient management program.

Chemical analysis of plant composition indicates chemicals or elements present in a crop at maturity or when it is harvested. For example, 1,250 lb of lint cotton contains approximately 125 lb of nitrogen (N), 20 lb of phosphorus (P), and 75 lb of potassium (K).

The essential question in fertilization is, "How much nutrient must be added to the soil as fertilizer for a given amount to be taken up by the growing plant?" The crop utilizes only a portion of the available nutrients in the soil. This means that more nutrients must be present than are removed by the crop. The amount added varies according to the level already present in the soil and the crop's need for the nutrient involved. The soil analysis is the starting point, since it measures the level or content presently in the soil.

The soil analysis along with the information provided in the information sheet, is interpreted and reported in terms of the nutrients needed to supplement those in the soil. With this information, producers can add sufficient nutrients for the correct balance to obtain high yields.

Limiting Factors

Crop yields are determined by a variety of factors including crop variety selection, available moisture, soil fertility, crop adaptation to the area, and the presence of diseases, insects, and weeds. The soil analysis and its interpretation deal only with the fertility level (plant nutrients) of the soil. Recommended fertilizer will provide sufficient nutrients for the best possible yields. Other factors of production or management may still cause low yields, even though nutrients are adequate.

Carryover

If yields are only partial in relation to a large amount of fertilizer applied, many of the nutrients are carried over for use by the next crop. It is this carryover, or residual effect, from one year to the next that makes heavy fertilizer applications practical in the face of other limits to yield.

Yields to Expect

A certain fertilizer application cannot be expected to produce a specific yield such as two bales of cotton or nine tons of hay. It is more realistic to assume that a balanced fertilizer program assures that nutrients are not the limiting factor in yields obtained. Research has shown that producers who use a balanced fertilizer program obtain consistently better yields than those who don't.

The Soil Analysis Report

After the soil is analyzed, fertility recommendations are made based on amounts of actual nutrients in the soil, not on the amount of any particular fertilizer or mixture. For example, if 100 lb of N were recommended, that amount could be supplied by approximately 300 lb of ammonium nitrate (33%N), 220 lb of urea (45%N), or 120 lb of anhydrous ammonia (82%N). Likewise, a recommendation of 60 lb of P205 per acre could be added as 133 lb of 45% triple superphosphate.

Fertilizer Labeling

Nitrogen is expressed on the elemental basis as "total nitrogen" (N). Phosphorus is expressed on the oxide basis as "available phosphoric acid" (P205). Potassium is expressed as "soluble potash" or potassium oxide (K20).

In reality, there is no P205 or K20 in fertilizers. Phosphorus exists most commonly as monocalcium phosphate, but also occurs as other calcium or ammonium phosphates. Potassium is ordinarily in the form of potassium chloride or sulfate. Furthermore, P205 and K20 are not absorbed by plants. Plant roots absorb most of their phosphorus in the form of orthophosphate ions, H_2PO_4-, and most of their potassium as potassium ions, $K+$. For these reasons, the elemental expression (N-P-K) is used in all of the recent research publications. Conversions from one form of P and K to another can be made using the following formulas.

%P=%P205 x0.437   %K=%K20x0.826

%P205=%Px2.29   %K20=%Kx1.21

Interpretation of the Soil Analysis Report

The soil analysis report contains two parts: characterization and fertility status of the soil, and fertility recommendations. Soil characterization (pH, texture, percent exchangeable sodium, percent organic matter, and salinity expressed as electrical conductivity) is explained in the report. The fertility status is reported as nutrients available to the plant.

The second part, fertility recommendation, contains the suggested amounts of fertilizer to apply. These amounts are based on the crop requirements, management practices affecting the crop (as shown in the information sheet), the present fertility level of the soil, and the yield goal desired by the producer. Special notification is given if the tests indicate that a salt or sodium hazard exists or if the information provided shows any other specific problems.

Soil amendments or treatments to reduce a sodium or salt hazard will be recommended if requested. In general, application of gypsum is suggested for reducing a sodium hazard, and leaching is recommended in most cases to lower salt content in the soil. Gypsum or leaching requirements are calculated and reported if requested.

Where to Get Soil Analyzed

There are many soil testing laboratories in New Mexico, Texas, Colorado, and Arizona. Basic soil testing packages vary in price and number of analyses. Many labs are participating in the Western Region Soil Testing Proficiency Program. Program participants share identical soils and compare results quarterly. This process assures the clients that the lab is striving for consistency and accuracy in lab analyses. Recommendations will undoubtedly vary from lab to lab. Often the best recommendation will come from the local Extension service. The choice of labs is at the client's discretion but should be based on report readability, result accuracy, turn-around time, and cost factors. New Mexico specialists can assist with many questions regarding plant health. Remember, a soil analysis is only as good as the soil sample taken.

Soil Analysis: Key to Nutrient Management Planning

Soil provides a reservoir of nutrients required by crops and also therefore for animals but not necessarily at optimum levels of immediate availability to plants. The purpose of soil analysis is to assess the adequacy, surplus or deficiency of available nutrients for crop growth and to monitor change brought about by farming practices. This information is needed for optimum production, to avoid transferring undesirable levels of some nutrients into the environment and to ensure a suitable nutrient content in crop products.

Farm assurance schemes, buyer's protocols and codes of practice are increasingly demanding more accurate fertiliser recommendations which must depend on the nutrient-supplying capacity of the soil. Regular soil analysis should be undertaken as a vital part of good management practice.

What is Measured?

pH, Phosphate, Potash & Magnesium

The "standard soil analysis package" measures soil acidity (pH) and estimates the plant-availableconcentrations of the major nutrients in the soil-phosphorus (P), potassium (K) and magnesium (Mg). Soil acidity is measured as pH (the concentration of H^+ ions) the scale running from pH 1 (very acid), through pH 7 (neutral) to pH 14 (very alkaline). The normal soil range is pH 4.5-8.5.

The total P and K content of a soil can be measured exactly but has little relevance to crop yield because only a relatively small proportion of the total P and K in soils is available to the plant. Soil analysis in the laboratory therefore uses chemical extractants to provide an estimate of the nutrient which would be available under field conditions. The results provide the best practical guide for determining P, K and Mg in the readily plant-available pool shown in the diagram. The methods of soil analysis used in the UK have been developed over many years and have been correlated to crop response on a wide range of soils in numerous field experiments. Other measurements of nutrient content or ratios may be made on soils, but unless there is dependable correlation with yields, they are of limited practical value. Calcium (Ca) and sodium (Na) can also be measured, but these are rarely determined.

Basal Nutrients in Soil

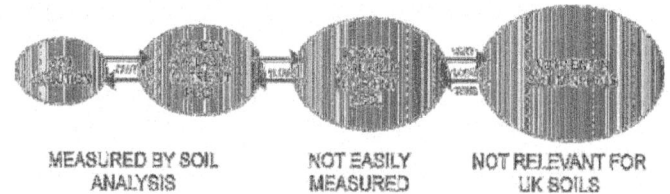

MEASURED BY SOIL   NOT EASILY   NOT RELEVANT FOR
ANALYSIS   MEASURED   UK SOILS

Fig. 1

Trace Elements

Of the elements required in small amounts by crops-the 'trace' elements or 'micronutrients'-some can be measured effectively in soil, namely boron (B), chloride (Cl), copper (Cu), molybdenum (Mo), cobalt (Co) and zinc (Zn).

Other elements which are not effectively assessed by soil measurements and need to be measured in herbage or in the animal include manganese (Mn), iodine (I) and selenium (Se).

Heavy Metals

In the UK there are statutory soil limits for 7 potentially toxic elements (PTEs) to ensure compliance with EU legislation when sewage sludge is to be applied. These are the 'heavy metals' cadmium (Cd), chromium (Cr), copper (Cu), lead (Pb), mercury (Hg), nickel (Ni) and zinc (Zn). The total (acid extractable) content of each of these in soil is analysed. The regulations also link the level of these elements to pH which affects their availability to plants. The limits for PTEs are defined in the Sludge (Use in Agriculture) Regulations 1989 and given in the Defra Soil Code. The Code advises the same limits be used to guide the application of other metal containing organic manures and wastes applied to land.

Nitrogen and Sulphur

Measuring the availability of soil nitrogen (N) and sulphur (S) can be useful but is more complex than for other nutrients; these are not covered in this leaflet.

Factors Affecting Soil Analysis

Nutrient values can vary as a result of a number of factors which are discussed below. Further study is needed to define the scale and cause of such variation.

Spatial Variation

Variation can exist over very short distances (less than 1 metre) and the number of cores that are needed to ensure that sampling is representative of an area will vary according to the scale of this variation. When sampling to obtain an average value for a field this is not so important because all the cores taken are bulked into a single sample; for a typically uniform field 25 cores are sufficient. However, local variation is important when grid sampling because each grid point is viewed as a separate sample. Normally, because the extent of the variation is unknown, at least 16 cores need to be taken on 1 metre grid-spacing around each grid point to provide a single bulked sample for analysis.

If it is planned to produce a 'map' of the nutrient variability within the field, then the grid points should be not more than 50 metres apart. Where considerable differences occur within a field, a single sample can be misleading because the averaging will disguise different treatment requirements. If different areas are known to vary they should be sampled separately. Knowledge of field amalgamations, soil type changes observed when ploughing or cultivating, visual crop growth difference across the field or any other specific information on field variation should be used to ensure that the area is sampled to best effect.

Temporal Variation

Variations in nutrient values have been observed at different times of the year and there is evidence to suggest that soil P, K and Mg values will be higher in the early spring than in the autumn as a result of chemical weathering over winter, biological activity and lack of uptake by growing crops

Moisture

If soils are dry at the time of sampling the analytical results can be affected and may appear a little lower for pH, P and possibly K. Movement and uptake of all nutrients will of course be restricted in very dry soils but this is a transient problem and does not reflect the normal availability of nutrients in the soil. At present it is not possible to quantify this effect in order to improve interpretation of analytical results.

Crop Removal

During periods of rapid growth crop uptake, especially of potassium, can be large and may deplete available soil nutrient levels for a short period until the nutrient status returns to equilibrium. This could affect results for some soils if samples are taken at such times.

Nutrients are returned to the soil in crop residues but will not be determined by analysis until the plant material is broken down. It has been suggested that sampling close to harvest may result in an under-estimation of true soil nutrient status.

Sampling Depth

Frequently there may be a gradient in nutrient level down the soil profile, usually declining with depth, reflecting the accumulation of nutrients in the plough layer. This is accentuated in minimal cultivation systems where phosphate and potash may be concentrated in the top 2 or 3 inches (5 to 8 cm). With continuous direct drilling there may be a large difference between the top 2 inches (5 cm) and

3 to 6 inch (8 to 15 cm) zones. For permanent grass where there is no soil disturbance, consistency of sampling depth of 3 inches (8 cm) is particularly important.

Previous Manuring

Applications of fertilisers and manures obviously have a major impact on measurement of soil nutrients. It is difficult to lay down rules as to how soon sampling should be undertaken after application. General guidelines are as follows:

> *8 weeks after P, K, Mg fertiliser application.12 weeks after slurry or manure application.12 months after lime spreading.*

Ploughing and cultivating help distribute nutrients from fertilisers and manures, and lime, throughout the depth of cultivation, but this takes time.

Sampling

Soil analysis data are only as good as the sample taken. A sample normally comprises around 1 kg (2 lbs) of soil which is taken to represent an entire area or field, which contains around 2,000 t of soil per hectare to a plough depth of 8 inches (20 cm). It is therefore imperative to obtain as representative a sample as possible or the results will not reflect the nutrient status accurately.

Rules of Sampling

1   Use of a suitable tool (cone auger, screw auger, corer etc) which facilitates and encourages the taking of more rather than fewer cores, of a uniform size and down to the full depth of sampling.

2   Use equipment and packaging that will not contaminate the sample. Galvanised sampling tools are unacceptable for trace element analysis.

3   Label samples clearly.

4   Sample to a consistent depth. Normal depth is 6 inches (15 cm) for arable soils, 3 inches (7.5 cm) for grassland.

5   Divide the field into areas which are as uniform as possible in soil type, past cropping, and manuring history and sample separately. Small areas of different soil e.g. wet, chalky, shallow, stony etc. should be excluded.

6   Avoid headlands, gateways, trees, mole hills, dung/urine patches, water troughs, areas where lime or manure has been dumped, old hedgerows/middens/ponds or any other irregular feature.

7   Discard stone and plant debris.

8   Take at least 25 cores from each area to be sampled and put them together to form a single representative sample. The numbers of cores should not be restricted simply because the container is full! Thoroughly mix all cores and take a sub-sample from this for despatch to the laboratory-this must be done carefully.

9   Ensure the sample represents the whole area. Sample on a W pattern over the field; for a regular shaped field this means 7 cores per leg of the "W".

10 In the case of grid sampling up to 16 cores are needed at each sampling point to obtain a representative sample. These should be taken in a regular pattern about 1 metre apart around the grid point. Grid points should be evenly spaced over the field and should not be more than 50 metres apart if a map is to be produced.

11 Sample at the same point in the rotation, before the crop which is most demanding or responsive to P and K. In descending order of importance these are: horticultural crops, vegetables, roots, pulses, spring sown combinable crops, winter sown combinable crops. For pH it is preferable to sample 12 months before a sensitive crop such as sugar beet or barley.

12 Sample at the same time of the year.

13 Avoid sampling under extremes of soil conditions e.g. waterlogged or very dry soil.

14 Do not sample within 8 weeks of fertilising, or within 12 weeks of manure or slurry application, for P, K and Mg analysis or sooner than 12 months after liming for pH analysis.

15 Maintain records and use the analytical results to develop nutrient management plans.

16 Where sampling to diagnose a crop problem take multi-cored samples (at least 16 cores per bulked sample) from areas of poor growth and separate samples from normal (good) areas.

The relative values between good and poor will be more informative than the actual values of the problem area.

Frequency of Sampling

Soil nutrient levels do not alter markedly over short periods of time unless major factors of supply or demand are introduced. There is therefore no point in incurring the additional cost of sampling more frequently than necessary, except perhaps as a check to ensure that earlier sampling has been carried out properly.

Additional sampling may be justified when there is a major change to husbandry practice-for instance alteration of cropping or manure policy. General guidelines for sampling frequency are as follows:-

Sample every

Permanent grass	7 years
Intensively used grassland	3-4 years
General arable cropping	4-5 years
Arable/grass systems	4-5 years
Field vegetables & horticulture	2-3 years

Analysis

There are a number of reasons why different results may be obtained for identical samples analysed by different laboratories. The most obvious reason is that different extractants may have been used. Variations in results can also occur if identical protocols are not followed-techniques of soil drying, grinding and sieving, reagent and equipment temperature, extractant concentrations, extraction shaking, stirring and filtration, and extraction time can all affect the result.

Results are usually reported in mg/l but where a laboratory measures soil by weight instead of volume, the units will be as mg/kg. This can give rise to differences on some soils-particularly those with high levels of organic matter.

If different laboratories are used for analyses these factors must be taken into account.

Interpretation of Soil pH

In England & Wales target pH (measured in water) is 6.5 for arable crops and 6.0 for grassland. Optimum range in Scotland where Scottish Agricultural College (SAC) measure pH in calcium chloride is 6.0-6.5 for arable crops and 5.7-6.2 for grassland.

In-field measurements using pH indicator on some soils where free chalk or lime particles exist may give lower values than laboratory results for the same field. This is because grinding the soil for laboratory analysis pulverises any chalk/lime particles and the pH as measured is increased.

Acidity below pH 6.0 will reduce the availability of nutrients, especially P. Soil analysis may overstate the availability of K on some alkaline soils where pH is over 7.0 and responses to added potash may occur at higher soil K levels than would be expected.

Availability of trace elements is radically affected by pH and the need for trace elements should only be assessed after any required amendment of acidity has been undertaken and has had time to take effect.

Interpretation of Soil P, K & Mg

Soil analysis provides an estimate of available P, K and Mg concentrations in soil to sampling depth-in practice this is equivalent to plough or cultivation depth because of the distribution of nutrients when the land is worked. Response experiments with different crop groups have provided the relationship between crop yield and soil nutrient concentration.

Normally, yields increase with increasing nutrient concentration to a maximum, beyond which there is no further benefit from additional nutrient. Below this value, which will vary with crop species, there is a yield penalty. Whilst soil analysis is not a precise guide, the lower the value the greater the risk of poor performance. To aid interpretation of the different concentrations of individual nutrients, Index or descriptive scales are used. These scales provide a general indication of the likely crop response and therefore a guide to the need for additional nutrient supplementation, as shown in the table.

Table 1: Crop response and soil analysis

		Yield response to added nutrient by	
Defra Index	SAC description	vegetable crops	arable crops & grass
0	Very low	highly likely	highly likely
1	Low	highly likely	probable
2	Moderate	likely	unlikely
3	High	possible	nil
4	High	unlikely	nil
5	High	nil	nil

Soil P, K & Mg Concentrations (mg/l) and Defra Index Scale

Note that the index is split in half for potassium only and described as 2- (or lower index 2) and 2+ (or upper index 2). In the past, index 2 was not divided in half for potassium but some soil reports used + and-signs to denote the extreme top and bottom 10% of each band; laboratories should no longer be using this convention.

Besides providing a basis to decide fertiliser quantities, soil analysis should also be used to monitor changes in fertility especially where there are uncertainties in the amounts of nutrient removed (e.g. with forage crops) and in the amounts of nutrients applied (e.g. with manures and slurries). For this purpose it is desirable to use the mg/l values not the index. However differences of less than 5 mg/l Olsen P, 25 mg/l K and 10 mg/l Mg should be ignored unless part of a sustained trend. Where accurate nutrient balance information is used in conjunction with regular soil analysis, it is important to recognise the possibilities of variation as discussed above.

Soil P, K and Mg Concentrations (mg/l) and SAC Descriptive Scale

The Scottish Agricultural College laboratory uses different extractants to those used in England and Wales and a descriptive rather than a numeric scale.

Table 2:

Description	Phosphorus	Potassium	Magnesium
	Modified Morgans extraction		
Very low	0-10	0-40	0-20
Low	10-25	40-75	20-60
Moderate	26-75	76-200	61-200
High	76-200	201-1000	201-1000
Excessively High	201-	1001-	1001-

Table 3: Relationship between Defra and SAC Scales

Defra Index	SAC description
0	Very low
1	Low
2	Moderate
3-7	High
8-9	Excessively high

Principles of P, K & Mg Manuring

The principle of manuring is to maintain plant-available soil nutrient levels within a target range depending upon crop rotation and soil type, by replacing nutrients removed. Soil analysis shows the status of soil nutrients relative to the target values and allows changes as a result of husbandry to be monitored. Where soils are below the target level, nutrient applications should provide more than is removed by the crop to ensure full yield response and to improve nutrient levels. Nutrient applications for soils above the target range may be reduced or omitted until the soil reserve approaches the target value. Additional nutrients may be applied before very responsive crops such as potatoes with the surplus balance being allowed for in subsequent less responsive crops. The overall nutrient balance in the rotation should be estimated and checked by regular soil analysis.

Sandy Soils

On true sand textured soils (not loamy sands or sandy loams) it is not practical to attempt to maintain soil K at 2- and the target K level should be adjusted to 100-120 mg/l (index 1+).

Chalk and Limestone Soils

At very high pH, potash additions are more likely to be converted to slowly available reserves in the soil and it will be more difficult to raise the soil K index. However soil K targets and replacement principles for application remain the same as for other soils.

Other Factors Affecting Interpretation

Soil structure is also very important because any restriction to root growth may restrict the plants ability to obtain an adequate nutrient supply despite a satisfactory value indicated by analysis. The remedy is not to apply more nutrient but to improve the soil structure.

Organic matter-The level of organic matter (humus) will also affect the availability of nutrients in a soil and regular addition of manures so that the physical conditions and biological activity is improved will increase the plant-available nutrients.

Stone content can also have a large effect upon nutrient supply. Very stony soils have little fine earth yet it is the nutrients in this fine earth fraction that are measured. In consequence it is advisable to maintain very stony soils at slightly higher levels of available P and K than are required on deep stone-free soils. Soil depth-Crops frequently access some nutrients from below sampling depth and from subsoil (especially potash). Soils with greater rooting depth potential, and therefore volume, provide larger total quantities of nutrient than shallow soils. It may be possible to manage deep, well textured soils at slightly lower concentrations than the stated targets given above. However shallow rooting crops which do not explore the full soil volume will still require normal nutrient concentrations.

Changes in Soil Nutrient Status

Changes in nutrient status relate to the balance between nutrients applied and removed or lost (little P, K or Mg is leached from most properly managed UK soils). Nutrient balances per hectare are normally small especially in comparison with the total quantity of soil nutrients per hectare. Where the balance is positive not all the residual phosphate or potash applied remains as available soil P and K and therefore only small changes in soil status should be expected. Large changes, where not related to very large nutrient balances, need to be investigated.

Improving Soils with Low Nutrient Status

Soil analysis measures the average nutrient concentration in the depth of soil sampled and theoretically a concentration of 1 mg/l to a depth of 10 cm in soil represents 1 kg/ha of (elemental) nutrient. In practice however, plants also obtain nutrients from below sampling depth and from the slowly available pool that is not measured by analysis. This often gives rise to some confusion especially as these sources are very variable depending on soil type and past fertilisation and manuring. It is therefore possible to have a soil at 125 mg/l of K which supplies a cereal crop with a peak uptake of 250 kg/ha K (300 kg/ha K_2O).

Conversely the addition of 100 kg/ha of phosphorus, potassium or magnesium (in fertiliser terms this represents 230 kg P_2O_5 120 kg K_2O and 170 kg MgO) will not result in an increase of 100 mg/l of readily plant-available soil P, K or Mg. This is because some of the nutrient applied will not remain in the readily available pool measured by analysis. Unfortunately the proportion remaining available and thus affecting soil analysis values will differ widely with different soils and conditions.

With present knowledge it is not possible to provide more than a very rough guide as to how soil values will change with additions (or removals) of nutrient. The following figures have been suggested but it is considered that the range in practice is even wider than indicated:

To increase soil Olsen P by 10mg/l requires 400-600 kg/ha P_2O_5

To increase plant-available soil K by 50mg/l requires 300-500kg/ha K_2O

More potash will be required on heavier soils where the clay type can make a difference.

Where either the P or K status needs to be increased, triple superphosphate and muriate of potash (KCI) respectively are cost-effective nutrient sources.

Treatment of High Fertility Soils

No additions of nutrient (fertilisers or manures) should be made to any soils above index 4.

Soils at index 3 should receive maintenance dressings of phosphate but no potash for most arable crops (except potatoes) or grassland.

Some clay soils contain very large reserves of potash which are not reflected in the level of available K shown by normal analysis. In these cases the amount of slowly available K released each season is sufficient to replace some or all of the potash removed in combinable crop rotations, especially where the straw in not removed (the higher demand of roots and cut grass normally requires some addition). The soil K value may not change over many years cropping even though no potash is used to replace that removed. Not all heavy soils have this ability and where less nutrient is applied than is removed, soil levels must be regularly monitored for any reduction in plant-available K.

Conclusion

Whilst soil analysis is not a perfect tool, it is the most effective and practical means of assessing soil fertility in respect of pH and plant-available P, K and Mg. Analytical data should be used in conjunction with other knowledge such as soil type, structure and crop offtake as the basis for deciding on fertiliser and manure use.

Analysis and Collection of Soil Samples

Back in the nineteenth century Edmund Locard developed the theory known as the "Locard's Exchange Principle." The theory in short states that when ever someone comes in contact with another object or person there is a minute exchange of particles that, in theory, can be traced back to a victim or suspect (Block 1999). In a crime context, that evidence can confirm or disprove a hypothesis of involvement between a suspect and a victim. Locard described these particles as dust or dirt but today it is understood to include all soil borne trace evidence. Trace evidence can include blood, hair, fibre, dirt, glass particles and any other minute particles at a crime scene.

It would be nice to think that this forward thinker, Locard, moved beyond his Principle to developed the idea of soil analysis but he didn't. His mentor and friend Hans Gross published articles almost simultaneously with Sir Arthur Conan Doyle, the author of the Sherlock Holmes Fictions, regarding soil analysis (Block, 1999)(Nickel & Fischer, 1999) To Edmund Locard's great disappointment both Arthur Conan Doyle and Hans Gross beat him to the punch on the evaluation of soil as a forensic tool. However, Locard's zeal for the use of trace evidence in forensic investigations led to a life long study of the classification and identification of soil samples (Block, 1999). Since its dawn in the late 1800's the analysis of soil has grown into a multidisciplinary field of forensic study. Modern analysis of soil may involve geologists, entomologists, toxicologists, biologists, botanists and a myriad of other experts.

Soil may be understood as just that: soil. As such it has a value in forensic studies. However, soil may also be understood as a repository for non soil contaminants which can yield valuable information about crimes. In forensic analysis materials can be grouped in several ways and each lab has its own way of subdividing these groups (Chayko & Gulliver 1999). In general soil borne materials can be considered organic or inorganic. Both types are found in soil. Further refining of these classifications where soil is concerned is to break the groups into mineral, biological or synthetic matter.

Soil is considered trace evidence. Soil is made up of disintegrated surface material which can be organic, mineral or synthetic. The ratio of the mineral content compared to other matter in the soil can be very site specific. The ratios of mineral, organic and synthetic matter can vary even with in a few feet (Chayko & Gulliver, 1999). Sandy soils look, feel and behave quite differently from clay soils or peaty soils. By profiling an array of characteristics of each soil, it is somtimes possible to attribute those characteristics to a specific location (Steck, 2004). Soil samples when properly taken can tell an investigator a lot about where a victim or suspect has been. Analysis of soil samples taken from vehicles can also tell an investigator about where a vehicle has been. Analysis of foot wear, clothing and tires can also place a suspect or victim in a particular location.

Collection of Samples

Collection of soil samples will depend on the circumstances of the crime. Indoor scenes will differ markedly from outdoor scenes in the type of evidence that can be recovered and the way in which these samples are collected.

At indoor scenes there may be footprints in soil or in dust. Samples made by footwear should be photographed to scale before being recovered. The particle samples can best be collected using a vacuum method. The samples can be vacuumed with a portable vacuum cleaner equipped with a special attachment. The attachment has a metal screen on which a filter paper is attached. The area is vacuumed and the filter is removed and labeled with the date, location, time and name of the technician who operated the vacuum. The vacuum must be thoroughly cleaned between samples. Cleaning can be fairly easily done with handhelds where the parts are easy to access. Reference samples from the surrounding area perhaps including flower gardens, points of entry and exit and alibi locations should also be taken.

Information on obtaining these specialized vacuum attachments can be obtained through Sirchie Laboratories Inc. in Youngsville N.C.

In the case of a break and enter or other crime at a home or business it is useful to know how the perpetrator entered the building. For example, if the perpetrator stood in a flower garden outside a window this indicates a stranger. If the perpetrator walked up the front lawn or to the back door this could indicate someone familiar with the property. Possibly the perpetrator knew no one was there being familiar with the owners habits. The soil on the shoes of a suspect (or left on a carpet or floor) can indicate the direction of travel and the mode of entry. This type of evidence is most useful when a suspect can be immediately identified before the soil is lost from the footwear.

Soil evidence from a victim or suspects clothing can indicate an association between victim and suspect. For instance, if a suspect lives in a particular neighborhood with a specific soil profile and this profile matches one found at a crime scene or on a victim, then one must suspect that the suspect left that sample at the scene.

"Double transfer" is even more convincing. If soil profiling to a suspects home territory is found on a victim or in their home and soil from the victims home territory is found on the suspects clothing or footwear the probability of the suspect having been at the crime scene increases dramatically. The mathematic probability of two matching profiles being found at both scenes by coincidence increases. For example the probability of transfer to one site occurring by coincidence might be 1 chance in 800. When double transfer has occurred the probability of that occurring by coincidence becomes 1 chance in 64,000 (Crocker, 1999).

Inorganic or manufactured matter found in the soil recovered from a suspect's footwear or clothing can be very site specific. Particles of glass, rubber and other industrial products can be used to link a suspect with a particular location. In the not-uncommon case of crimes at industrial locations this can be particularly useful.

How to Collect Samples

As already mention in the indoor scene vacuuming can be used to collect samples. The methods of collection will vary from scene to scene. In a break and enter where house plants are found upset a sample of the soil may be used to link the suspect to a scene. Soil from the garden or yard can also be used to track the suspect's direction of travel and point of entry.

Most garden soils are unique. The gardener in charge will have added some favorite materials (sometimes specific to particular plants) in order to improve the garden. The materials tend to be different in every garden. Some gardeners might use compost while others might prefer ammonium pellets, sand, peat or wood chips.

Samples taken from the interior of a vehicle can indicate many things. A soil sample from the gas or brake pedal of a vehicle can link a suspect to a location. Soil from a trunk or backseat can indicate digging. When graves are dug to hide bodies the various levels of the soil are disturbed.

If the suspect lays a shovel on the back seat or in the trunk the soil from the lower levels of the grave may be deposited on the seat or in the truck. Soil from tools such as shovels should be preserved in the state it was found. The entire tool should be packaged in protective material then enclosed in plastic. Finally the entire tool should be placed in a wooden or cardboard box and transported to the lab.

Sampling methods may need to be adapted when the scene is outdoors. If the challenge is the recovery of remains, soil samples should be taken at regular intervals up to 100 yards from the gravesite or point of recovery (Saferstein, 2004). Usually a grid search is set up and samples can be taken from each square of the grid and labeled as to which grid it was taken from.

About a tablespoon of soil should be enough for most modern tests. Usually only the surface soil needs to be sampled. The exception to this is in the case of a buried body, In this case soil samples should be taken at regular intervals as the remains are exposed. After the remains are removed from the gravesite the bottom of the grave should also be sampled. It is important to note that a new shovel, spoon or other scoop must be used for each grid and sample. Should the same implement is used serially, uncleaned to recover sample then Locard's Principle is at work again and cross contamination will render your samples useless.

Samples should be placed in plastic vials for transport. If it is not possible to transport the samples immediately they should be allowed to air dry before transport.

A special note for gravesites is that if insect evidence is present the vials should be labeled in pencil rather than pen. The specimens will likely be preserved in alcohol which if it leaks onto the label will destroy ink from pens. Without the label as to the time and place of collection your sample is forensically worthless.

Outdoor scenes often involve vehicles. There are several ways of collecting soil samples from vehicles. Vehicles involved in accidents will sometimes leave lumps of soil from under the wheel wells and fenders on the road way. These lumps should be collected intact and wrapped in protective material to minimize bumping during transport. The purpose behind this is to preserve the layers that have built up to create this lump of soil.

A soil analyst can read these layers and know where this vehicle has been. The analyst may also be able to match the layers in a lump of soil to a particular vehicle. I can't help but think this would have been very useful in tracking Ted Bundy or Henry Lee Lucas in their cross county travels.

Clothing and footwear should be collected intact. No attempt should be made to remove soil from clothing, footwear or tires. If these items can be removed intact they should be placed in a paper bag or enclosed in a druggist's fold and then placed in a paper bag. Care must be taken so that the paper bag is protected so that evidence is not lost through holes in the bag. Some evidence can be placed in plastic bags but is must be completely dry. Wet samples placed in plastic with quickly degrade and rot and become useless. Plastic does not "breath" to allow passage of moisture and air. Paper on the other hand will allow moisture to escape preventing rot.

Lumps of soil stained with blood, semen or other biological samples should be collected intact and transported to the lab as dry samples. Any samples containing suspected biological material such as blood, flesh, semen or hair must be clearly labeled so that the analyst at the lab can take precautions to preserve this material. Biological specimens in soil must never be heat dried.

The only time a sample should not be allowed to dry is when insect evidence such as maggots are present. In the case of maggots there is a very specific way of handling this evidence. Samples containing maggots can be placed in aluminum foil with a small piece beef liver and placed in a plastic container. There must be air available in the container. These samples must be sent to the lab immediately. The specimens can be placed in a thermally protected case such as a cooler. Attempts should be made to protect the samples from extremes of heat and cold. Never use dry ice with biological or entomological (insect) evidence.

In the Laboratory

In the lab items from the victim and suspect should be examined separately. Ideally the items and samples from the victim and suspect should be processed in different rooms. The personnel handling these sample should be assigned to one or the other. If this is not possible then the person handling the material should take extreme care to avoid cross contamination of the samples. The laboratory staff will evaluate the soil samples in several ways. First the mineral content will be tested. Some experienced analysts can moisten a sample and feel the soil. Based on feel alone they can tell the ratio of mineral and organic content. Microscopic examination of soil samples will subsequently reveal the type and nature of the mineral, biological and synthetic content of a sample.

Such talented analysts are not always available. Then, there are several standard testes which can identify the type and origin of the sample. The first and probably the most common is the density gradient tube. Two different liquids are added to a glass tube in various ratios. Each ratio represents a different density. The soil sample is poured into the tube. When the various particles reach a level in the liquid where their density is equal to the liquid the particles become suspended. This creates a unique profile of bands in the tube which can be matched to other samples.

The samples may also be tested using heat to test the point at which the sample will undergo an exothermic reaction or an endothermic reaction. The sample is heated in a special furnace to various temperatures. In an exothermic reaction the sample essentially burns and releases heat. In an endothermic reaction the sample will absorb the heat. Each sample from different locations will have these reactions at different temperatures according to the mineral, biological and synthetic content.

Electron microscopes can be used to reveal the crystalline structures of minerals and synthetic material in a sample of soil.

Nuclear Resonencing and Mass Spectometry are also methods which may be used in the Laboratory.

It is important that the crime investigators and the laboratory analyst communicate properly. Perhaps this happens best when the laboratory head is knowledgeable about investigations and is the contact person for investigators. The head can be briefed on the questions and issues of the case and then direct the laboratory personnel as to the direction of the inquiry. The investigators may want to know if sample #12 and sample #76 are similar. The laboratory head will then choose appropriate methods which express the similarity between samples. Once the questions and procedures are chosen everything depends on the integrity of the sample. If they have been collected with care and documented amply then the results from the laboratory can be trusted (Steck, 2004).

Conclusion

Take samples of all the soil in and around a crime scene. Place a reference sample in a plastic vial and label it with the date the time the investigators name and the case number. Reference samples are samples of soil from places the suspect or victim may have picked up soil. The reference sample can also be from sites the suspect may have been. Include the location and the distance from the focal point of the crime when labeling samples.

Reference samples should be about a tablespoon or so taken from less than a 1/2 an inch (about 1 cm). The exception is in cases of burial where samples should be taken every 1 inch (every 3cm). With gravesites the bottom of the grave should be sampled after the removal of the remains. Package and label all samples carefully. Keep in mind that the analysts in the laboratory do not have firsthand knowledge of the scene and the circumstances of collection. Try to provide a detailed description of the scene and the circumstances of the crime which the laboratory staff can use to determine what type of analysis is most appropriate.

Chapter 5

Soil pH and Electrical Conductivity

Purpose

This manual has been designed as a reference source for county Extension laboratories offering soil pH and/or electrical conductivity tests to their clients. This manual if followed, will assist county faculty in assuring that these laboratory measurements are done correctly with high quality assurance.

Soil pH and its Uses

Soil pH measurement is useful because it is a predictor of various chemical activities within the soil. As such, it is also a useful tool in making management decisions concerning the type of plants suitable for location, the possible need to modify soil pH (either up or down), and a rough indicator of the plant availability of nutrients in the soil.

Aluminum

Aluminum in the soil can adversely affect plants if aluminum occurs in certain forms and its activity is elevated sufficiently. As the activity of aluminum increases, the soil becomes more acidic and soil pH decreases. If the pH is low enough and aluminum is present in sufficient quantity, plants may be stunted or lost due to aluminum toxicity. In most Florida sandy soils, there is danger of aluminum toxicity when soil pH is below 5.0.

The occurrence of aluminum toxicity (at acidic pHs) decreases somewhat as one travels from the Florida panhandle and down the peninsula. The reason for this apparent change is found in changes in the soil constituents. Aluminum levels tend to decrease and the aluminum is found in different minerals in the soil as on travels south in Florida.

Organic soils have low total aluminum. Low soil pH in these soils does not pose a threat from aluminum toxicity in Florida.

Nitrogen Fixing Microbes

Legume plants have a helpful relationship with selected soil microbes. These microbes convert nitrogen gas from the atmosphere to forms useful to the plant for growth and improved yield. In turn, the microbes are supplied nutrients and carbohydrates from the plant. This mutual, beneficial existence is termed symbiosis. Agronomic examples of this symbiotic relationship are alfalfa, peanut, and soybean.

Soil pH directly affects the activity of these microbes. Research, conducted in the absence of aluminum toxicity, has shown that once the soil pH has decreased to 4.7 or lower, the ability of the microbes to convert nitrogen is greatly reduced. If aluminum is present, then both the microbial symbiotic activity and the normal metabolism of the plant are adversely affected.

Solubility of Plant Nutrients

Soil pH directly affects the solubility of many of the nutrients in the soil needed for proper plant growth and development. These chemical reactions are complex and have often been generalized with charts that over simplify chemical conditions in specific soils. One should be careful in using these pH nutrient charts when dealing with soils in Florida. They may be misleading.

As soil pH decreases, nutrients, such as phosphorus, usually decrease in plant availability because of precipitate reactions with iron and aluminum. However, plants can affect their micro-environment and are often found to grow well over a range of soil pH. This range of successful growth is often as great as 1 to 2 pH units. In general, many plants will do well in a soil pH range of 5.5 to 7.5. Specific plants, such as azalea or pine seedlings, actually require low soil pH. Such plants are often iron in efficient, meaning that they require low soil pH to aid in the up take of iron from the soil.

As soil pH increases above 6.5, manganese, a micronutrient, may become limiting to plant growth. Phosphorus and micronutrients such as copper and zinc also decrease in their plant availability at high pH. Soils composed of limestone (such as those in the Dade County area) have a high native soil pH of about 8.3. Plants grow in these high pH soils, but nutrient deficiencies are common.

Optimum pH Ranges for Plants

As stated above, most plants do well over a range of soil pH values. This point bears repeating because the best management of soil pH is often to do nothing. Because many plants directly modify the chemical environment around their roots, nutrient limitations are not found, plant production is not adversely affected, and visual stress symptoms are not observed.

The University of Florida, IFAS, Extension Soil Testing Laboratory (ESTL) uses the IFAS Standardized Fertilization Recommendation System. This system contains all of the IFAS approved fertilizer and liming recommendations (Hanlon et al., 1990). Specific pH values are cited in this system as target pHs. A target pH is a soil pH, within the optimum pH range, that is used for the calculation of lime rates. A target pH is usually selected such that the adverse effects of aluminum toxicity are avoided, and so that nutrient availability for that crop will be adequate. A target pH is not the only pH at which the crop will do well.

Liming of Soils

If the soil pH is greater than 0.2 units below the target pH, the ESTL will complete an additional test, the Adams-Evans Buffer. This test has been specifically designed for the sandy soils found in the southern United States. Furthermore, current lime recommendations have been calibrated for Florida conditions using both the soil pH and the Adams-Evans Buffer.

Soil pH is a measure of the active acidity, that portion of the hydrogen ions that is active in the soil solution. Soil pH does not measure the reserve acidity. Reserve acidity is that portion of hydrogen and other acid-contributing ions that are sorbed on soil particles. Usually, the reserve acidity is much greater than the active acidity. The Adams-Evans test is designed to measure this reserve acidity. Together, the target pH, the soil pH, and the Adams-Evans test can be used to determine the amount of lime required to adjust the soil pH from its current reading to the target pH.

To see the effect of reserve acidity on lime recommendations. Soil samples from the indicated counties were all selected to have the same soil pH of 5.7. However, the amount of lime in pounds/acre or pounds per 1,000 sq ft to raise the soil pH to a target pH of 6.0 (columns 3 and 4) or to a target pH of 6.5 (columns 5 and 6) varies widely.

Unfortunately, the Adams-Evans test is rather complex and contains chemicals that must be handled and disposed of as hazardous waste. For this reason, county Extension laboratories should not offer the Adams-Evans test. However, local knowledge of the county and its soils can help considerably with local lime recommendations. It is recommended that liming recommendations made at the county laboratory be verified by submitting a small percentage (e.g., 5%) of soil samples to the ESTL. Comparison of the local recommendations with those from the ESTL will allow county Extension faculty to calibrate their recommendations.

Acidifying Soil

In some situations, it may be desirable to acidify the soil, that is, to lower soil pH. If the high soil pH is a natural condition, there is little that can be done to lower soil pH permanently. Treatment with sulfur will lower the pH for a few weeks, but the pH will eventually increase.

In landscaping, it is often better to select plants which are adapted to the natural soil pH range, rather than to use plants which will need constant soil pH maintenance and usually look unhealthy even after this extra effort.

In situations where the high soil pH condition was created by human activity, for instance overliming, it is often feasible to lower the soil pH with one or two applications of sulfur. In such cases, the amount of overliming is relatively small, say 1 ton/acre (46 lb/1000 sq ft) and can be treated successfully. In naturally occurring, high pH soils, the effective lime equivalent will usually be over 100 tons/acre.

Unfortunately, there is no soil test available to assist in determining the amounts of sulfur (S) needed to reduce soil pH. If plants are actively growing, agricultural sulfur treatment should be restricted to a maximum of 300 lb S/acre (7 lb S/1,000 sq ft). Damage may still occur at this rate if the S is allowed to remain in contact with the foliage.

Multiple treatments during the growing season should be done with caution; that is, allowing enough time (usually about 1 month) for the previous S treatment to react with the soil. Sulfur added to the soil must undergo oxidation by soil microbes resulting in the production of hydrogen ions and sulfate. Since microbial action controls the effectiveness of the treatment, warm moist soil conditions are preferable to dry or cool conditions. Treatment of the soil with gypsum, which is calcium sulfate, will not change soil pH because gypsum does not contribute any acidity to the soil. It is a pH-neutral salt.

pH Electrodes

The electrode and its reference electrode may be combined into a single combination electrode. However, for ease of understanding, this discussion will deal with separate electrodes. All electrode parts are similar, whether they are used as two separate electrodes or as a single combination electrode.

The glass ball at the end of the pH electrode is composed of a special glass which has specific surface properties. The ball should not be touched nor allowed to dry out. Inspection of this part should include insuring that the ball has not been cracked, that it is full of solution, and that there is a thin wire extending into the upper portion of the ball within the electrode.

If the electrode has been inverted, the ball may not contain any solution. Gently tapping the electrode side with one's fingers while holding it vertical with the ball down is usually enough to displace the entrapped air with solution.

Most manufacturers include instructions for maintaining (rejuvenating) the electrode if the ball has been allowed to dry out. Rejuvenation often calls for strong acids which present a personal safety hazard. County faculty may wish to replace electrodes rather than attempting rejuvenation.

Inspection of the electrode wire should not reveal any loose connections or breaks in the insulation. There are no serviceable parts on the pH electrode.

Reference Electrode

The electrode contains a fibrous or ceramic chip which allows the internal solution to slowly flow out from the body of the electrode into the external solution of the sample which is being measured. This leaking results in a completed electrical circuit. Inspection should include checking that the chip is in place and is relatively clean of debris from past use. Often the chip will be white or light gray in colour, indicating that it is still functioning. If the chip is a different (darker) colour, the electrode may not function properly. Use of a small amount of abrasion may restore the chip, but too much abrasion may damage the electrode.

Lastly, the barrel of the electrode often contains a filler port near the top of the electrode. Most manufacturers recommend that the covering be removed from this port so that pressures do not develop within the operating electrode. The solution, often a concentrated solution of potassium chloride, should be kept within 1 inch of the filler port. This level is sufficient to insure that the internal solution level is higher than the sample solution being analyzed. This difference permits flow out from the electrode through the chip.

A relatively new type of electrode is constructed with an internal fluid which is a gel. This type of electrode does not contain a filler port because it can not be serviced by the user.

pH Meter

The pH meter is a sensitive electronic device, often designed with printed circuits making it quite reliable. However, improper handling (bumps, hot dash-board storage, or solution spills) can damage the meter. The meter is designed to measure changes of millivolts between the reference and pH electrodes. To understand the sensitivity needed to measure millivolts, human neck pains are often induced by 7 to 20 millivolts.

Most problems associated with reading pH usually originate with faulty electrodes. Additionally, if the meter contains batteries, they should be checked before each use. Low or inadequate power supply will result in inaccurate pH readings.

In any case, unless the instrument completely fails, it will always give a pH reading whether it is functioning correctly or not! It is the responsibility of the operator to detect problems and to report accurate pH results.

Electrical Conductivity and its Uses

The Electrical Conductivity (EC) of a solution is a measure of the ability of the solution to conduct electricity. The EC is reported in either millimhos per centimeter or the equivalent decisiemens per meter. When ions (salts) are present, the EC of the solution increases. If no salts are present, then the EC is low indicating that the solution does not conduct electricity well.

The EC indicates the presence or absence of salts, but does not indicate which salts might be present. For example, the EC of a soil sample might be considered relatively high. No indication from the EC test is available to determine if this condition was from irrigation with salty water or if the field had been recently fertilized and the elevated EC is from the soluble fertilizer salts. To determine the source of the salts in a sample, further chemical tests must be performed.

Soluble Salts vs. EC

Prior to 1989, the ESTL reported soluble salts values. Soluble salts is an older term and is derived from EC measurements. Unfortunately, there are a number of assumptions which have to be made before soluble salts can be calculated. Conversion factors (EC to soluble salts) were different across the United States, ranging from 600 to 700. Florida used a conversion factor of 700. These assumptions, and their introduced errors, are avoided by reporting the actual EC measurement.

The EC can also be directly used in the body of literature which is continuously growing concerning plant productivity under the effects of salinity. The older term soluble salts should be avoided by county Extension laboratories.

Interpretation of Electrical Conductivity of Irrigation Water

Frequent use of irrigation water will directly influence the salts in the soil profile. Salts are influenced by factors such as rainfall content and timing, internal soil drainage, and irrigation practices. Usually, rainfall contains low amounts of salts and acts to dilute salts that are present in the soil. If the rainfall is of sufficient volume or duration, and the soil has internal drainage, the added rainfall is enough to leach salts from the soil.

During drying conditions, water is lost from the soil due to evaporation, and salts are effectively concentrated. If irrigation water contains appreciable salts, then intensive management is required to produce healthy plants. Therefore, EC measurement of the irrigation water source is an excellent management decision. Table below has been developed to assist in making a decision to use a water source.

Table 1: Classification of irrigation water by electrical conductivity.

Class of Water	Specific Conductance dS/m
Excellent	<0.25
Good	0.25 to 0.75
Permissible	0.76 to 2.00
Doubtful	2.00 to 3.00
Unsuitable	>3.00

Interpretation of Electrical Conductivity of Soils

In actuality, the interpretation of EC of a soil or media must be made considering the plant(s) to be grown. The EC of the soil has little direct detrimental effect on sandy mineral soils or on media. However, EC directly affects plants growing in the soil or media. The impact of EC on plants is also directly affected by water management.

Salt Index Use and Calculation

As EC increases, more attention to water management is needed to prevent salinity from adversely affecting plants. The Extension Soil Testing Laboratory uses a 2:1 solution:soil ratio with which to determine EC. Many states use a saturated paste extract. This saturated paste method is more time consuming than the 2:1 extraction, and results in inadequate amounts of solution in Florida's sandy soils. The conversion from the 2:1 extraction result to the saturated paste result, termed salt index, is easy and accurate.

EC (salt index) = EC (2:1) x 8.

In general, when the soil EC (2:1, water:soil) exceeds 0.25 dS/m (or 0.25 x 8 = 2.0 dS/m salt index), many plants experience stress due to salts. Other plants (e.g., bermudagrass) are quite tolerant to salts. Due to this species-dependent effect of EC, a listing of the effects of increasing EC on selected plants has been compiled (Hanlon et al., 1993).

EC Probe and Meter

There are a wide variety of EC probes and meters marketed. Those instruments using probes that are placed in the soil or media *in situ* (directly in the soil without taking a sample) are not considered in the following discussion. Since EC is a measurement of the conductivity of the soil solution, the measurement should be made under a controlled mixture of solution to soil (2:1). These conditions usually do not exist when direct reading instruments are used.

Electrical Conductivity Probe

The probe consists of a tube, usually of plastic, into which electrodes have been installed. Two common electrode arrangements are: 1) two plates; or 2) a rod located concentrically in a ring. In either case, the electrodes are held a specific distance apart. The gap between the electrodes is filled with the water sample or filtered solution from a soil sample, either by filling a reservoir or by placing the probe in the solution.

When the probe is immersed in the solution, ions contained in the solution will permit electrical flow from one electrode to another. If a large number of ions are present (salty conditions), then the EC of the sample will be higher than a sample with low number of ions (low salts).

Some older instruments require that a mixture of soil and solution be packed into the electrode receptacle. While readings from these older instruments may be adequate, accurate results are harder to obtain because readings are affected by the sample packing method.

Inspection of the probe should include insuring that the probe is clean and free of debris, and that all electrical connections are in place. Some probes may experience a buildup of corrosion on the electrodes with time. Indication of corrosion, if not directly visible, is indicated by the constant need to reset the instrument because of drifting, often in one direction as corrosion progresses.

Electrical Conductivity Meter

Meters for EC are extremely reliable. Meters from the 1950s which use a "cat's eye" tube are still in use. Newer meters using digital displays are often susceptible to failure of one or more segments of the digital display, usually related to corrosion within the instrument.

As with failures associated with pH meters, EC meters will always provide an EC, unless the problem results in total instrument failure. It falls upon the operator to insure that the reported EC is an accurate measurement of the conductivity of the sample.

Laboratory Procedures

The following procedures are used at the ESTL and are based on good laboratory procedures with sufficient quality control measures to ensure that pH and EC readings are accurate and reliable. While equipment may vary among County Extension Laboratories, procedures should be developed that directly parallel those used by the ESTL.

Soil Scooping Technique

The ESTL uses a scoop (that is, a volume\ measurement) for both pH and EC determinations. The scoop is a plastic and metal device which may be obtained from a commercial manufacturer (e.g., Custom Laboratories, Orange City, FL). Alternately, the scoop may be constructed from locally available materials, such as measuring spoons or coffee scoops. The intent is to use a consistent volume of soil and water. The soil scooping technique requires practice, despite its unsophisticated appearance. The technique depends upon uniform actions by the technician from sample to sample to produce consistent packing of soil into the scoop. This consistency can be directly measured by repeatedly scooping the same soil and weighing each scoop. Weights should be uniform within each scoop (volume) and soil-sample combination. Weights will vary from soil to soil, especially when there is a noticeable difference in soil texture.

Sample Handling and Preparation

The sample should be air-dried and passed through a 2-mm sieve before scooping.

Procedure

1. Dip the scoop into the center of the soil sample and fill the scoop with a twisting motion so that extra soil is mounded above the rim of the scoop. Do not press the scoop or force the soil against the side of the container (Jones, 1980).

2. Strike the handle near the scoop three times with a plastic rod to settle soil particles.

3. Level the scoop with the plastic rod. Strike off all excess soil above the rim of the scoop in a single stroke so that the soil is not compacted into the scoop.

Estimation of Soil Texture

Knowledge of soil texture is useful in the recommendation of lime. Local soil conditions may be quite uniform so that little differences may be found throughout the county.

However, such uniformity is expected to be rare in most counties. After the soil has been air-dried and sieved, a small amount of dry soil should be moistened and rubbed between the forefinger and thumb. An estimation of the texture is made by comparing the"feel" of the sample to that of a set of soil samples of known texture.

Soil pH (2:1 V/V)

This procedure uses a 20-cc (~25- g) soil scoop and 40 mL of pure water to obtain a 2:1 water-to-soil ratio. Most problems with this procedure are associated with the glass or calomel electrodes. However, sample pH may also be affected by contaminated water, by microbial activity if samples are allowed to sit for several hours before determining pH, or by improper scooping techniques.

1. **Standard Solutions:** Obtain commercial standard solutions of pH 4.00, 7.00, and 10.00.

   Sample Handling and Preparation: The soil sample should be air-dried and passed through a 2-mm sieve. Irrigation water samples require no preparation.

   Procedure: 1. One scoop of soil to a 3-oz plastic cup using a 20-cc (~25- g) scoop.

2. Add 40 mL of pure water to each cup using an automatic pipette or suitable volumetric container. Stir with a glass rod and let the sample stand for 30 min.

3. Standardize the pH meter (see following section, pH Meter).

4. Stir the sample again immediately before measuring the soil pH. Do not place the electrode(s) directly in the sand layer at the bottom of the container. The electrodes should be positioned in the solution just above the sand layer.

5. Record pH to the nearest 0.1 pH unit (suggested format of XX.X).

Electrical Conductivity

This test, often called "Soluble Salts," requires that 20 cc (~25 g) of soil be mixed with 40 mL of pure water, resulting in a water:soil ratio of 2:1. The 4-hr equilibration period provides time for some slowly-soluble constituents to approach solution equilibrium. Little error results from much longer equilibration times, but shorter time periods might introduce inconsistent results for some samples.

Standards

A solution of 0.005 N KCl has an electrical conductivity of 720 ± 1 dS/m (mmho/cm) at 25°C. Use a commercially prepared solution as the reference solution.

Sample Handling and Preparation

• The soil sample should be air-dried and passed through a 2-mm sieve. Irrigation water samples require no preparation.

Procedure

1. Place 20 cc (~25 g) of soil in a plastic 3-oz cup.

2. Add 40 mL of pure water, stir, and allow to stand for 4 h.

3. Without stirring the sample, filter the solution through a Whatman No. 41 (11 cm) paper and collect the extract in a funnel tube or other suitable container. The intent is to remove the soil and other debris from the solution.

4. Standardize the conductivity meter.

5. Move the probe up and down in the solution several times to dislodge any bubbles on the electrode surfaces. Measure the electrical conductivity of the extract contained in the funnel tube.

6. Rinse the interior and exterior of the probe with pure water between samples. Remove any excess water from the exterior of the probe by blotting with a tissue.

7. Record all meter readings as displayed. Note that some meters use a floating-point display while others use a reading which depends upon a switch setting.

pH Meter

Instructions for the proper use of a pH meter, including electrode care and instrument-specific settings, are given in this subsection. While instrument settings may vary among meters, daily operation of pH meters is relatively standard. Most problems with pH instruments originate within the electrode(s) or in the electrical connections to the instrument. Prevention of early electrode deterioration is best accomplished by following the information in sections B and C that follow. Use of only one standard buffer solution is not adequate for proper calibration. Do not reuse a buffer solution which has been left out overnight or has been used to "store" electrode(s). It is advisable to maintain a supply of a reference soil sample to be used as a check sample with every set of samples. Early detection of problems is the best method of avoiding inaccurate results.

Calibration of Electrode(s)

1. Rinse electrode(s) with pure water and blot dry with a tissue or clean paper towel.

2. Using commercially prepared buffer solutions of pH 4.00, 7.00, and 10.0, pour approximately 30 mL into labeled, 3-oz. plastic cups. Do not reuse buffer solutions in which electrode(s) have been immersed for daily storage.

3. Set the meter to operate in the "pH" mode.

4. Place the electrode(s) into new buffer solution (pH 7.00) and allow the electrode(s) to equilibrate. A stable reading should be obtained within about 30 seconds. Readings which are not stable after 1 minute indicate that the electrode(s) may be malfunctioning.

5. Using the Calibration" (often labelled "Standard") knob, adjust the pH-meter display to read 7.00. Do not move (or press) either the "Temperaturer" or "Slope" adjustments at this time.

6. Place the meter in the "Standby" mode. Some meters do not have a standby mode.

7. Rinse the electrode(s) into a waste cup with pure water and blot dry as above.

8. Place the electrode(s) in new buffer solution (pH 4.00) and allow the electrode(s) to equilibrate with the buffer solution.

9. Set the meter to operate in the "pH" mode.

10. Using the "Temperature" knob (sometimes labelled "Slope"), adjust the pH-meter display to read 4.00. Do not readjust the "Calibration" setting at this time.

11. Repeat Steps 6 through 9 using the pH 10.0 buffer solution. No settings should be changed as the pH 10.0 buffer is being read.

12. Repeat Steps 3 through 11 until readings of 7.00, 4.00, and 10.0 are obtained without adjusting the instrument. These standards should be set ±0.05.

13. Using a reference soil sample, read the pH and determine if the pH reading agrees with the "known" reading of the reference soil. Use of a reference soil sample, a soil that can be analyzed with every sample set, is strongly recommended. Use of a reference soil is an excellent quality assurance measure and verifies that the electrode(s)/meter are functioning correctly in a soil solution and not just in standard buffers. The reference soil sample should be read once about 20 samples. If the reading drifts by more than ±0.2, then restandardize the meter.

14. Standardize the instrument according to the above procedure after every 50 samples. If the instrument has shown drift of 0.05 pH units or more, reread the last few samples to verify the accuracy of the recorded readings.

Daily Electrode Storage

1. Place the meter in the "Standby" mode.

2. Wash the electrode(s) with pure water.

3. Immerse the electrode(s) in pH 7.00 buffer solution.

Electrode Servicing

1. If the pH reading drifts, replace the glass electrode. Wait until the reading from the new electrode is stable. Usually, new electrode(s) should be kept in pH 7.00 buffer solution overnight before use.

2. After use, wash the electrode(s) with pure water.

3. Immerse the electrode(s) in pH 7.00 buffer solution.

Electrical Conductivity Meter

Conductivity measurements must be made on solution samples only. The most common problem with conductivity meters concerns failure of the electrodes (mounted within the hollow plastic probe) to make proper contact with the solution.

The probe must be kept clean by adequately flushing the interior of the probe with pure water. When analyzing a solution, insure that the solution covers the bottom 3 to 5 cm (1 to 1.5 inches) of the probe. Agitation of the probe in the unknown and standard solutions is required to obtain reproducible results. Agitation insures complete wetting of the internal electrodes by removing any air bubbles using the solution of interest.

Calibration

1. Turn ON the "Supply" or "Power" switch.

2. Turn OFF the "Temperature Correction" switch (back panel).

3. Set the "Scale" wafer switch, if any, to read mmho/cm.

4. Place the conductivity probe into a standard solution of 0.005 M KCl. Agitate the probe using an up-and-down movement to create better solution to probe contact.

5. After agitation, the meter should read 0.72 ± 0.04 mmho/cm (0.72 ± 0.04 dS/m).

6. Agitate the probe again and reread the standard solution. Both the first and second readings should be the same value if the probe is in good contact with the solution.

7. Wash the probe with pure water on both inner (inject wash water through the hole at the top of the probe) and outer surfaces. Blot the probe dry. Do NOT rub the outer surface of the probe with the tissue paper.

8. Read the reference soil sample and verify that the current reading agrees with the known value for the reference soil.

9. Occasionally, "EEE" will appear on the instrument display indicating that the sample reading exceeds the current scale setting. Move the "Scale" wafer switch to the next highest position and reread the sample. This switch changes the reading by factors of 10. Note the correct decimal reading.

10. Record the entire meter reading including location of the decimal point.

11. Return the "Scale" wafer switch to the original position before reading the next sample.

Soil Electrical Conductivity Variability

Soil electrical conductivity (EC) is a property of soil that is determined by standardized measures of soil conductance (resistance[1]) by the distance and cross sectional area through which a current travels. Traditionally, soil paste EC has been used to assess soil salinity (Rhoades et al., 1989), but now commercial devices are available to rapidly and economically measure and map bulk soil EC across agricultural fields.

The Veris® 3100 (Veris Technologies, Salina, Kansas) measures EC with a system of coulters that are in direct contact with the soil. The EM38 (Geonics, Limited, Mississauga, Ontario, Canada) induces a current into the soil with one coil and determines conductivity by measuring the resulting secondary current with another coil. Both sensors have been demonstrated to give similar results (Suddeth et al., 1999)

The movement of electrons through bulk soil is complex. Electrons may travel through soil water in macropores, along the surfaces of soil minerals (i.e. exchangeable ions), and through alternating layers of particles and solution (Rhoades et al., 1989). Therefore, multiple factors contribute to soil EC variability, including factors that affect the amount and connectivity of soil water (e.g. bulk density, structure, water potential, precipitation, timing of measurement), soil aggregation (e.g. cementing agents such as clay and organic matter, soil structure), electrolytes in soil water (e.g. salinity, exchangeable ions, soil water content, soil temperature), and the conductivity of the mineral phase (e.g. types and quantity of minerals, degree of isomorphic substitution, exchangeable ions).

Despite the multiple causes of EC variability, bulk soil EC measurements have been related to individual factors that limit soil use and productivity such as salinity (De Jong et al., 1979; Rhoades and Corwin, 1981), clay content at a depth of 15-m in New Wales, Australia ($r^2 = 0.78$; Williams and Hoey, 1987), depth of sand deposition along the Missouri River ($r^2 = 0.73$-0.94; Kitchen et al, 1996), depth to claypan in Missouri ($r^2 = 0.73$; Doolittle et al., 1994), and soil moisture content ($r^2 = 0.96$; Kachanoski et al., 1988).

If soil EC maps have utility in production agriculture, 1) EC must be spatially structured, 2) spatial patterns must have temporal stability, and 3) EC must be related to factors of agronomic importance. The objective of this research was to determine the nature and causes of EC variability for several fields in Kentucky. Geostatistical analyses were conducted to examine the spatial and temporal variability of EC variability. Transect studies were conducted to determine the causes of EC variability.

Methods

Site Description

This research was conducted in a field in Hardin Co, KY (Field 1), and three fields in Shelby Co, KY (Field 2, Field 3, and Field 4). Field 1 consists of the Vertrees series (Fine, mixed, mesic Typic Paleudalfs), which formed in residuum from limestone, the Nolin series (Fine-silty, mixed, mesic Dystric Fluventic Eutrochrepts), which formed in mixed alluvium, and the Crider series (Fine-silty, mixed, active, mesic Typic Hapludalfs), which formed in loess over limestone residuum. Field 2 consists of the Nicholson series (Fine-silty, mixed, mesic Typic Fragiudalfs) and the Lowell series (Fine, mixed mesic Typic Hapludalfs), both of which formed in residuum from limestone and shale and have a loess cap. Field 3 has both the Nicholson series and the Shelbyville series (Fine-silty, mixed mesic Mollic Hapludalfs), which formed in loess over residuum from limestone. Field 4 contains the Cynthiana series (clayey, mixed, mesic Lithic Hapludalfs) and Faywood series (fine, mixed, active, mesic Typic Hapludalfs), both which are shallow to bedrock and formed in residuum from limestone.

Soil EC Data Collection

A Veris® 3100 Soil EC Mapping System was used to measure soil EC. The sensor consists of six coulters, two of which introduce an electrical potential into the soil. The remaining four coulters are spaced to measure EC over two approximate depths, 0-30.5-cm ($EC_{30.5}$) and 0-91.5-cm ($EC_{91.5}$). When used in conjunction with a DGPS receiver, EC data can be geo-referenced to create a map.

Transect EC Measurements

Transects were selected from Field 1, Field 3, and Field 4. At selected points on each transect, several measurements were taken in addition to EC. Volumetric water content to a depth of 12-cm was determined with the HydroSense™ (Decagon Devices, Inc, Pullman,

Washington), which uses transmission line oscillation. Depth to a clay increase was assessed using the "texture by feel" method. Penetrometer resistance was measured on all Field 1 transects. Depth measurements to fragipan (Field 3) and to bedrock (Field 4) where measured only in the fields where these factors affected crop production. Soil EC values for each point on the transect were determined by driving directly over the sampling point, stopping, and recording the $EC_{30.5}$ and $EC_{91.5}$ values. In addition, the transect EC values in Field 3 were measured on two consecutive days.

Whole Field EC Measurements

Whole field EC data were collected for a location in Hardin Co. (Field 1) on three dates: Oct. 8, 1999, March 7, 2000, and May 5, 2000. A second location in Shelby Co. (Field 2) was measured once on July 7, 1999. The fields were traversed with approximately 7.5-m between passes on each date. Field 3 was traversed in the north-south direction and the east-west direction. Data were recorded every second, and groundspeed was maintained at approximately 10.5 km hr^{-1}. Soil samples were collected for the top 15-cm using a 30.5-m grid pattern. Soil analyses, including pH, buffer pH, organic matter content, and Mehlich III extractable P, K, Ca, Mg, and Zn, were performed by the Department of Regulatory Services, at the University of Kentucky. At each sample site, all EC values falling within a 4.6-m radius were averaged and related to $EC_{30.5}$ and $EC_{91.5}$ using simple linear regression.

Results and Discussion

Nature of the Variability

Day-to-day variability was greater for $EC_{30.5}$ than for $EC_{91.5}$. While collecting EC data, extreme values were encountered on occasion, as can be seen by the high value for May 4th $EC_{91.5}$. The small-scale temporal variability also gives an indication of measurement error. The relative nugget variances for both depths also give an indication of measurement error ($EC_{30.5}$, 46%; $EC_{91.5}$, 27%) and were large.

The larger scale temporal variability reflects changes in EC associated with different environmental conditions. While the EC values were substantially lower during the drought of 1999 (October 8th, 1999) than in the spring of 2000, the general spatial patterns in EC were similar across all three dates.

Both $EC_{30.5}$ and $EC_{91.5}$ tended to be higher on the Vertrees series, which is located in the lower right, lower left, and upper right regions of Field 1. The values tended to be lower for the Nolin series, which are mainly located in the depressions. The remainder of the field was the Crider series. Soil EC was a good indicator of soil type for this field. The Vertrees soils have a red, Bt horizon near the surface, which increases conductivity. The EC values for the Vertrees series were not as high on October 8th, 1999 (during the drought) as on the later two dates. This suggests that soil moisture enhances the conductivity of clay.

This did not, however, greatly change the overall appearance of the maps on the different dates because each was based on an equal number of observations in each category rather than equal sizes of mapping intervals. Date of the measurement did, however, change the spatial statistics of the data. Anisotropic behavior was not apparent on October 8th, 1999, but it was on the latter two dates. The dark diagonal lines indicate the direction of the anisotropic axis with of minimum spatial variability (here the northwest-southeast direction). Orthogonal to this direction indicates the direction with the maximum spatial variability.

Anisotropy was present in this field because the Vertrees soil, which occurred in southwest, northeast, and southeast regions of the field, had very large EC values on the latter two dates when soil moisture was greater. Therefore, variability was much greater in the southwest-northeast direction. In all cases, the anisotropy does not seem to be a great issue within the first 50-m. Anisotropy was less important with $EC_{91.5}$. This may be because at greater depths, the soil was more uniform in clay and moisture content. The impact of anisotropy in this field would depend upon the depth of interest, whether the data would be used for interpolation, and the sampling interval.

Chapter 6

Soil Moisture Measurement Instrumentation

Introduction

Moisture content of the soil is a major factor determining plant growth, especially in irrigated systems. Currently there are many and varied methods for determining soil water content on a volume basis (q_v, m^3m^{-3}) or a tension basis (kPa or bar) as described by Gardener.

The basic objective of irrigation scheduling is to minimise water stress of the plant, that of over irrigation, and under irrigation. The manager aims to manipulate the biological process of cell elongation and cell reproduction for improved plant yield and maximum use of available effluent.

In optimising plant cell reproduction and growth (cell expansion), the ability to monitor the soil moisture content is the principal facet of developing good water management programs. A tendency to over or under-irrigate results due to the absence of information about the soil moisture status down the soil profile. The result of over irrigation is poor utilisation of natural rainfall because of high surface run off, and production problems associated with excessively wet soil such as waterlogging, leading to recharge of underlying aquifers, leaching of nutrients, increased incidence of plant disease and reduced daily water use. The reduced daily water use of plants increases the area of irrigated land required to dispose of a given volume of water increasing the capital cost of land based waste water disposal systems.

The decline of soil water content will result in a decrease of photosynthesis and cell expansion of the plant. Under-irrigation will result in stress being placed upon the plant root water uptake mechanism to maintain transpiration rates. A subsequent reduction in daily water use and cell production will occur with decreasing soil moisture content. Ludlow et al. showed that stem elongation rate declined (in *Cajanus cajan*) at a linear rate after 10% of available water was utilised by the plant until elongation was 40% of the maximum rate as the plants approached wilting point.

To develop an irrigation scheduling program the basic requirement is the ability to regularly obtain objective data. The ability to accurately measure soil water content, plant size and condition is an integral mechanism in the process of developing an irrigation scheduling program that allows a better understanding of plant and soil water relations. From this basis, an understanding of plant agronomy is developed with an appropriate computer interface giving the manager a better working knowledge of what is happening to the applied irrigation and its relation to plant water use and soil moisture status.

Instruments Available for Objective Measurement of Soil Moisture Content

Objective soil moisture measurement can be undertaken with simple tools, such as a shovel, or complex tools that record measurement of soil moisture on a volumetric basis. The method of measurement is simply a device allowing moisture determination in an objective fashion. It is important that measurements are made regularly and recorded systematically to allow improvement in irrigation scheduling and soil/plant management decisions.

There are different methods available for development of an irrigation scheduling program using different tools to collect relevant information and present the data to the irrigation manager.

The Neutron Probe (NP)

An established technique that is used extensively throughout Australia by farmers, consultants and researchers. The technique is based on the measurement of fast moving neutrons (generated from an Americium 241/Beryllium source) that are slowed (thermalised) in the soil by an elastic collision with existing Hydrogen particles in the soil. Hydrogen (H^+) is present in the soil as a constituent of:

1. soil Organic matter
2. soil clay minerals
3. water.

Water is the only form of H^+ that will change from measurement to measurement. Therefore any change in the counts recorded by the NP is due to a change in the moisture with an increase in counts relating to an increase in moisture content.

In the field aluminium tubes are inserted into the soil and stopped to minimise water entry. Readings are taken at depths down the profile (e.g. 20, 30, 40, 50, 60, 70, 80, 100 and 120 cm) with a sixteen second count. The three aluminium tubes are then averaged to counter the effect of spatial variability reducing the value of the measured moisture content data.

Measurements are taken two to four times a week and information is down-loaded to a personal computer for interpretation. Use of the NP technique for vadose (unsaturated) zone monitoring has been employed to determine contaminate leak detection along specific transport pathways, and to monitor land disposal of effluent with the NP technique.

Time-Domain Reflectometry

Determines the apparent dielectric (Ka) of the soil matrix and this is empirically related to the volumetric soil moisture content. The method is quick, relatively independent of soil type, non destructive, suited for surface and profile measurements, and allows repeatable in situ measurement. The TDR is a portable unit that can be carried allowing point soil moisture measurements or linked to a multiplexer to measure an array of buried waveguides.

The moisture content determined by the TDR is the average moisture along the length of the waveguides. Therefore, to measure at depth of 20 cm, waveguides are placed in the soil horizontally at that depth. If 30 cm waveguides are placed vertically into the soil, the moisture content determined by the TDR will be the integrated moisture content from the soil surface to a depth of 30 cm.

The technique is based upon cable testing technology, with a broad-band Electromagnetic step pulse generated and propagated along a coaxial cable (Fig. 1.). At the end of the cable stainless steel rods (waveguides) are inserted into the ground. The time of travel of the EM wave is determined by the apparent dielectric (Ka) of the medium (in this case soil). Water with a high dielectric (Ka 80), compared to soil (Ka 3 to 5) and air (Ka = 1), dominates the measured Ka. Thus, if the soil is saturated the Ka is high (due to the presence of increased water) and the travel time of the EM wave along the waveguides is long. If the soil is dry the travel time along the waveguides is short and the Ka is therefore low. Eq. 1. shows the relationship of Ka to travel time (Dt).

$$Ka = (cDt\,/2L)^2 \qquad \textbf{(eq. 1)}$$

Where c is the velocity of light (3×10^8 ms^{-1}) and L is the length of the wave guide (m). Topp et al. empirically related Ka to qv via third order polynomial and this equation (eq. 2.) is the basis for soil moisture measurements at present.

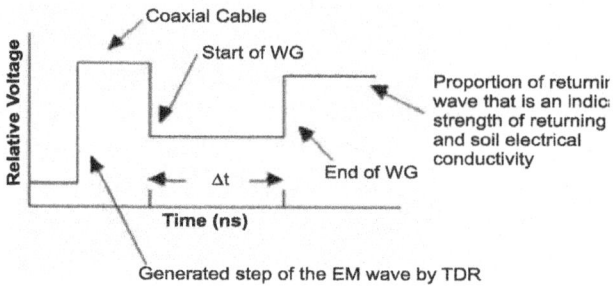

Figure 1. Schematic diagram of an electromagnetic wave generated by a step pulse TDR system as it travels along the coaxial cable and down the waveguides into the soil.

$$\theta v = -5.3\times10^{-2} + 2.92\times10^{-2}Ka - 5.5\times10^{-4}Ka^2 + 4.3\times10^{-1}$$ (**eq. 2**)

Further calibration is required for soil high in organic matter and other materials such as grain. In the field waveguides (stainless steel) are of two forms being either balanced (two-wire) or unbalanced (three-wire) as shown in Fig. 2. Generally, two wire probes are used for portable measurement and the three wire probes for permanently placed waveguides.

Effective length of waveguides (and therefore the depth of measurement) will be determined by the power of the step pulse generated by the TDR, the soil type (heavy clay attenuates the wave more so than lighter soil types) and the moisture content of the soil.

Waveguides of length 2 m have been successfully used to measure moisture content in Australian soil. However, in wet heavy clay soil waveguide length has sometimes been reduced to as little as 30 cm. This current problem is being rectified by increasing the power and stability of the EM wave and by coating waveguides with thin cover of a low dielectric material.

This will ensure that a percentage of the wave will travel the length of the waveguides and be reflected allowing determination of ?t. Importantly, the attenuation of the EM wave in conducting soil (soil with a high electrical conductivity) will allow the TDR technique to independently measure moisture content and bulk soil electrical conductivity.

This is important for the measurement of solute travel. Research is increasing in developing further applications of TDR such as surface measurements, profile measurements, long range multiplexing of waveguides and solute transport determination. This technique will be more widely used in the future by research and irrigation managers.

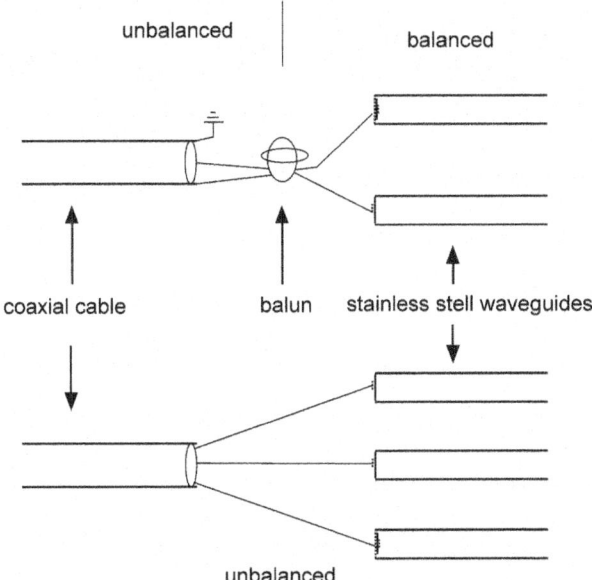

Figure 2. Schematic diagram illustrating the connection of the coaxial cable (unbalanced signal) to the stainless steel waveguides through, a) a balun for the two-wire system, and b) direct to the waveguides in the three-wire system.

Tensiometers

Portable and stationary tensiometers measure the soil moisture content as a tension or pressure ranging from 0 to -100 kPa, (0 to -1 bar). Tensiometers fundamentally act in a similar fashion to a plant root measuring the force that plants have to exert to obtain moisture from the soil. As the soil dries the water is lost from the tensiometer via a ceramic cup. The loss of water creates a vacuum in the tensiometer and is reported as a pressure reading, the drier the soil the higher the pressure reading, (noting -0.1 bar is considered field capacity and -15 bar wilting point).

Tensiometers may be placed permanently in the soil giving an analogue or digital output. Logging of tensiometers is possible via transducers and a communication cable back to a computer or datalogger. Portable tensiometers allow greater freedom of sampling giving relatively quick readings of soil moisture tension.

Tensiometers can take time to equilibrate especially in heavier soil types and this should be accounted for in determining an irrigation scheduling regime. Tensiometers must be installed correctly and well maintained to operate accurately and the practical limit for reliable readings generally -800 kPa (-0.8 bar).

Frequency Domain (Capacitance)

The capacitance technique is similar to that of TDR in that the apparent (Ka) dielectric of the soil is measured and empirically related to the moisture content (qv). A high frequency transistor oscillator (150 Mhz) operates with the soil (dielectric) forming part of an ideal capacitor as shown in eq. 3.

$$C = Ke_o \, A/s \qquad \qquad \text{(eq. 3)}$$

where the dielectric (K) is related to the capacitance (C) via the relationship of the total electrode area (A) and spacing of the electrodes (s), noting that (e_o) the permittivity of free space is constant.

In a field situation the design of the capacitance probe is not ideal with two annular rings (electrodes) placed in a plastic access tube in the soil. The measured area is now removed from between the electrodes to outside the access tube as shown in Fig. 3. Thus, in a field situation the measured capacitance (C) is determined as:

$$C = gK \qquad \qquad \text{(eq. 4)}$$

where the C is related to K via a geometrical constant (g). g depends upon electrode spacing, area and orientation of the electrodes in the soil and e_o.

Measurement is undertaken by either lowering a sensor into the access tube or placing an array of sensors into the access tubes and logging the output frequency. The measured (angular) frequency is related to the soil moisture content via a non-linear calibration. Measurement of absolute moisture content is dependant on soil type and bulk density.

The potential for capacitance based soil moisture determination is good. However, development is required to determine the actual measurement area of the probe and its spatial sensitivity to change in moisture content.

Further, the calibration of the technique *in situ* needs to more fully understood to allow universal use and the effect of electrical conductivity, temperature and acid soil on measured frequency has not been fully studie.

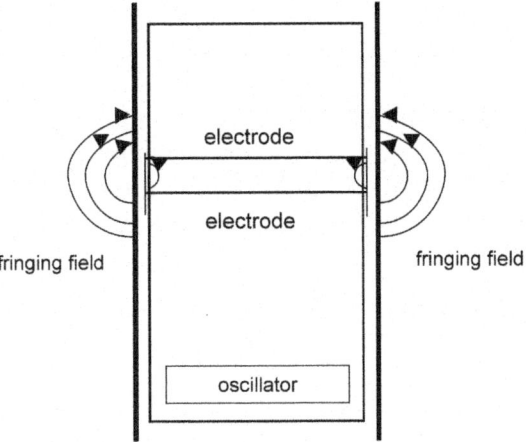

Figure 3. Schematic diagram of capacitance probe in an access tube (after White and Zegelin)

Electrical Resistance (Gypsum)Blocks

Electrodes are embedded in a porous (gypsum) block and placed in the soil at different depths in the root zone. The water in the soil will reach an equilibrium with the water in the gypsum block and the electrical resistance is then determined and related to moisture content as a tension (kPa or bars).

Gypsum blocks do not measure the moisture content at low potential (from 0 to -100 kPa) well. The operating range is suited from about -100 kPa to -1500 kPa (as the soil dries). Gypsum blocks will dissolve over a period of time (with the rate of dissolution increasing in sodic soil) generally lasting for two to three seasons in good conditions.

Large errors, up to 100%, can occur due to: slow equilibrium of blocks with the actual soil potential; the dependence of resistance on the block temperature; effect of hysteresis on calibration of block (if undertaken to improve accuracy) and actual contact with the soil; and, blocked pores by fine material (e.g. silt or clay particles).

Electrical resistance is a useful indicator of the soil moisture content in respect to root conditions such as: plentiful water; good growing conditions; approaching water stress; and water stressed plants.

The need to determine the moisture status of the soil is a critical factor influencing plant production. Correct irrigation scheduling can control the soil moisture status reducing through-drainage and maintaining optimum levels of soil water for maximum plant growth. To implement a reliable and accurate irrigation scheduling regime regular, objective soil moisture readings are essential. There are different tools available for obtaining soil moisture content including NP, TDR, tension and capacitance techniques. The choice of instrumentation will be determined by the form of information required by the operator, the soil type, relative cost, reliability and ease of use in the field.

Some Diseases of the Soil

Soil Erosion

Perhaps the most widespread and the most important disease of the soil at the present time is soil erosion, a phase of infertility to which great attention is now being paid.

Soil erosion in the very mild form of denudation has been in operation since the beginning of time. It is one of the normal operations of Nature going on everywhere. The minute mineral particles which result from the decay of rocks find their way sooner or later to the ocean, but many may linger on the way, often for centuries, in the form of one of the constituents of fertile fields. This phenomenon can be observed in any river valley. The fringes of the catchment area are frequently uncultivated hills, through the thin soils of which the underlying rocks protrude. These are constantly weathered and in the process yield a continuous supply of minute mineral fragments in all stages of decomposition.

The slow rotting of exposed rock surfaces is only one of the forms of decay: the surfaces not exposed are also subject to change. The covering of soil is no protection to these underlying strata, but rather the reverse, because the soil water, containing carbon dioxide in solution, is constantly disintegrating the parent rock, first producing subsoil and then actual soil. In this way the constant supply of minerals — like phosphates, potash, and the trace elements needed by crops and livestock — are automatically transferred to the surface soil from the great mineral reservoir of the primary and secondary rocks. Simultaneously with these disintegration processes the normal decay of animal and vegetable remains on the surface of the soil is giving rise to the formation of humus.

All these processes combine to start up denudation. The fine soil particles of mineral origin, often mixed with fragments of humus, are gradually removed by rain, wind, snow, or ice to lower regions.

Ultimately the rich valley lands are reached, where the accumulations may be many feet in thickness. One of the main duties of the streams and rivers which drain the valley is to transport these soil particles into the sea, where fresh land can be laid down. The process looked at as a whole is nothing more than Nature's method of the rotation, not of the crop, but of the soil itself. When the time comes for the new land to be enclosed and brought into cultivation, agriculture is born again. Such operations are well seen in England in Holbeach Marsh and similar areas round the Wash. From the time of the Romans to the present day new areas of fertile soil, which now fetch £100 an acre or even more, have been recreated from the uplands by the Welland, the Nene, and the Ouse. All this fertile land, perhaps the most valuable in England, is the result of two of the most widespread processes in Nature — weathering and denudation.

But Nature has devised a most effective brake. The nature of this retarding mechanism is of supreme importance, because it provides the key to the solution of the problem of soil erosion. Nature's control of the rate of denudation is to create the compound soil particle. The fragments of mineral matter derived from the weathering of rocks are combined by means of the specks of glue-like organic matter supplied mostly by the dead bodies of the soil bacteria which live on humus; as in a building made of bricks, some suitable cementing material is needed before the fragments of mineral matter in the soil can cohere. There must be sufficient of this cement of the right type always ready, so that when the mineral fragments come together a piece of glue is there at hand of a size corresponding to the minute areas of contact. This involves the constant production of large quantities of this bacterial cement. Provided, however, that we keep up the bacterial population of the land in any catchment area, the supplies of glue for making new compound soil particles and for repairing the old ones will be assured.

It will be seen from this how fundamentally important is the role of humus. It is the humus which feeds the bacterial life, which, so to say, glues the soil together and makes it effective. If the supply of glue is allowed to fall into arrears, the compound soil particles will soon lie about in ruins and so provide more raw material for speeding up the process of denudation. The mineral particles are thereby released and ready for their final journey by water to the sea to form new soil, or by wind to form a new dust bowl and so begin a new desert.

It is when the tempo of denudation is vastly accelerated by human agencies that a perfectly harmless natural process becomes transformed into a definite disease of the soil. The condition known as soil erosion — a man-made disease— is then established. *It is, however, always preceded by infertility:* the inefficient, overworked, dying soil is at once removed by the operations of Nature and hustled towards the ocean, so that new land can be created and the rugged individualists — the bandits of agriculture — whose cursed thirst for profit is at the root of the mischief can be given a second chance. Nature is anxious to make a new and better start and naturally has no patience with the inefficient. Perhaps when the time comes for a new essay in farming, mankind will have learnt the great lesson — how to subordinate the profit motive to the sacred duty of handing over unimpaired to the next generation the heritage of a fertile soil. Soil erosion is nothing less than the outward and visible sign of the complete failure of a farming policy. The root causes of this failure are to be found in ourselves.

The damage already done by soil erosion all over the world, looked at in the mass, is very great and is rapidly increasing. The regional contributions to this destruction, however, vary widely. In some areas like north-western Europe, where most of the agricultural land is under a permanent or temporary cover crop (in the shape of grass or leys) and there is still a large area of woodland and forest, soil erosion is a minor factor in agriculture. In other regions like parts of North America, Africa, Australia, New Zealand, and the countries bordering the Mediterranean, where extensive deforestation has been practiced and where almost uninterrupted cultivation has been the rule, large tracts of land once fertile have been almost completely destroyed.

The United States of America is perhaps the only country where anything in the nature of an accurate estimate of the damage done by erosion has been made. Theodore Roosevelt first warned the country as to its national importance. Then came the Great War with its high prices, which encouraged the wasteful exploitation of soil fertility on an unprecedented scale. A period of financial depression, a series of droughts and dust storms, emphasized the urgency of the salvage of agriculture. During Franklin Roosevelt's presidency soil conservation became a political and social problem of the first importance. In 1937 the condition and needs of the agricultural land of the United States of America were appraised. No less than 253,000,000 acres, or 61 per cent of the total area under crops, had either been completely or partly destroyed or had lost most of its fertility. Only 161,000,000 acres, or 39 per cent of the cultivated area, could be safely farmed by present methods. In less than a century the United States has, therefore, lost nearly three-fifths of its agricultural capital. If the whole of the potential resources of the country could be utilized and the best possible practices introduced everywhere, about 447,466,000 acres could be brought into use — an area actually greater than the present crop land of 415,334,931 acres. The position, therefore, is not hopeless. It will, however, be very difficult, very expensive, and very time-consuming to restore the vast areas of eroded land even if money is no object and large amounts of manure are used and green-manure crops are ploughed under.

Such, in this great country, are the results of misuse of the land. The causes of this misuse include lack of individual knowledge of soil fertility on the part of the pioneers and their descendants; the traditional attitude which regarded the land as a source of profit; defects in farming systems, in tenancy, and finance — most mortgages contain no provisions for the maintenance of fertility; instability of agricultural production as carried out by millions of individuals, prices, and income, in contrast to industrial production carried on by a few large corporations. The need for maintaining a correct relation between industrial and agricultural production, so that both can develop in full swing on the basis of abundance, has only recently been understood. The country was so vast, its agricultural resources were so immense, that the profit seekers could operate undisturbed until soil fertility — the country's capital — began to vanish at an alarming rate.

The resources of the Government are now being called up to put the land in order. The magnitude of the effort, the mobilization of all available knowledge, the practical steps that are being taken to save what is left of the soil of the country and to help Nature to repair the damage already done are graphically set out in *Soils and Men,* the Year Book of the United States Department of Agriculture of 1938. This is perhaps the best local account of soil erosion which has yet appeared. The progress that has been made in recent years can be followed in *Soil Conservation,* a monthly periodical issued by the Soil Conservation Service of the United States Department of Agriculture, Washington, D.C.

The rapid exploitation of Africa was soon followed by soil erosion. In South Africa, a pastoral country, some of the best grazing areas are already semi-desert. The Orange Free State in 1879 was covered with rich grass, interspersed with reedy pools, where now only useless gullies are found. Towards the end of the nineteenth century, it began to be realized all over South Africa that serious overstocking was taking place. In 1928 the Drought Investigation Commission reported that soil erosion was extending rapidly over many parts of the Union and that the eroded material was silting up reservoirs and rivers and causing a marked decrease in the underground water supplies. The cause of erosion was considered to be the reduction of vegetal cover brought about by incorrect veldt management — the concentration of stock in kraals, overstocking, and indiscriminate burning to obtain fresh autumn or winter grazing.

In Basutoland, a normally well watered country, soil erosion is now the most immediately pressing administrative problem. The pressure of population has brought large areas under the plough and has intensified overstocking on the remaining pasture. In Kenya the soil erosion problem has become serious during the last ten years, both in the native reserves and in the European areas. In the former, wealth depends on the possession of large flocks and herds; barter is carried on in terms of livestock; the bride price is almost universally paid in anima's s; numbers rather than quality are the rule.

The natural consequence is overstocking, over-grazing, and the destruction of the natural covering of the soil. Soil erosion is the inevitable result. In the European areas, erosion is caused by long and continuous over-cropping without the adoption of measures to prevent the loss of soil and to maintain the humus content. Locusts have of late been responsible for greatly accelerated erosion; examples are to be seen when the combined effect of locusts and goats has resulted in the loss of a foot of surface soil in a single rainy season.

The countries bordering the Mediterranean provide striking examples of soil erosion, accompanied by the formation of deserts which are considered to be due to one main cause — the slow and continuous deforestation of the last 3,000 years. Originally well wooded, no forests are to be found in the Mediterranean region proper. Most of the original soil has been washed away by the sudden winter torrents. In North Africa the fertile cornfields which existed in Roman times are now desert.

Ferrari in his book on woods and pastures refers to the changes in the soil and climate of Persia after its numerous and majestic parks were destroyed; the soil was transformed into sand; the climate became arid and suffocating; springs first decreased and then disappeared. Similar changes took place in Egypt when the forests were devastated; a decrease in rainfall and in soil fertility was accompanied by loss of uniformity in the climate. Palestine was once covered with valuable forests and fertile pastures and possessed a cool and moderate climate;

today its mountains are denuded, its rivers are almost dry, and crop production is reduced to a minimum.

The above examples indicate the wide extent of soil erosion, the very serious damage that is being done, and the fundamental cause of the trouble — misuse of the land, resulting in the destruction of the compound soil particles. In dealing with the remedies which have been suggested and which are now being tried out, it is essential to envisage the real nature of the problem. It is nothing less than the repair of Nature's drainage system — the river — and of Nature's method of providing the countryside with a regular water supply. The catchment area of the river is the natural unit in erosion control. In devising this control we must restore the efficiency of the catchment area as a drain and also as a natural storage of water. Once this is accomplished, we shall hear very little about soil erosion.

Japan provides perhaps the best example of the control of soil erosion in a country with torrential rains, highly erodible soils, and a topography which renders the retention of the soil on steep slopes very difficult. Here erosion has been effectively held in check by methods adopted regardless of cost, for the reason that the alternative to their execution would be national disaster. The great danger from soil erosion in Japan is the deposition of soil debris from the steep mountain slopes on the rice fields below.

The texture of the rice soils must be maintained so that the fields will hold water and allow of the minimum of through drainage. If such areas become covered with a deep layer of permeable soil, brought down by erosion from the hillsides, they would no longer hold water and rice cultivation — the mainstay of Japan's food supply — would be out of the question. For this reason the country has spent as much as ten times the capital value of eroding land on soil conservation work, mainly as an insurance for saving the valuable rice lands below. Thus, in 1925 the Tokyo Forestry Board spent 453 yen (£45) per acre in anti-erosion measures on a forest area valued at 40 yen per acre in order to save rice fields lower down valued at 240 to 300 yen per acre.

The dangers from erosion have been recognized in Japan for centuries and an exemplary technique has been developed for preventing them. It is now a definite part of national policy to maintain the upper regions of each catchment area under forest as the most economical and effective method of controlling flood waters and insuring the production of rice in the valleys. For many years erosion control measures have formed an important item in the national budget.

According to Lowdermilk, erosion control in Japan is like a game of chess. The forest engineer, after studying his eroding valley, makes his first move, locating and building one or more check dams. He waits to see what Nature's response is. This determines the next move which may be another dam or two, an increase in the former dam, or the construction of retaining side walls. After another pause for observation a further move is made and so on until erosion is checkmated. The operation of natural forces, such as sedimentation and re-vegetation, are guided and used to the best advantage to keep down costs and to obtain practical results. *No more is attempted than Nature has already done in the region.* By 1929 nearly 2,000,000 hectares of protection forests were used in erosion control. These forest areas do more than control erosion. They help the soil to absorb and retain large volumes of rain water and to release it slowly to the rivers and springs.

China, on the other hand, presents a very striking example of the evils which result from the inability of the administration to deal with the whole of a great drainage area as one unit. On the slopes of the upper reaches of the Yellow River extensive soil erosion is constantly going on. Every year the river transports over 2,000,000,000 tons of soil, sufficient to raise an area of 400 square miles by five feet. This is provided by the easily erodible loess soils of the upper reaches of the catchment area. Some of the mud is deposited in the river bed lower down, so that the embankments which contain the stream have constantly to be raised. Periodically the great river wins in this unequal contest and destructive inundations result.

The labour expended on the embankments is lost, because the nature of the erosion problem as a whole has not been grasped, and the area drained by the Yellow River has not been studied and dealt with as a single organism. The difficulty now is the over-popuration of the upper reaches of the catchment area, which prevents afforestation and laying down to grass. Had the Chinese maintained effective control of the upper reaches — the real cause of the trouble — the erosion problem in all probability would have been solved long ago at a lesser cost in labour than that which has been devoted to the embankment of the river.

China, unfortunately, does not stand alone in this matter. A number of other rivers, like the Mississippi, are suffering from overwork, followed by periodical floods as the result of the growth of soil erosion in the upper reaches.

Although the damage done by uncontrolled erosion all over the world is very great and the case for action needs no argument, nevertheless there is one factor on the credit side which has been overlooked. A considerable amount of new soil is being constantly produced by natural weathering agencies from the subsoil and the parent rock. This, when suitably conserved, will soon re-create large stretches of valuable land. One of the best regions for the study of this question is the black cotton soil of Central India which overlies the basalt.

Here, although erosion is continuous, the soil does not often disappear altogether, for the reason that, as the upper layers are removed by rain, fresh soil is re-formed from below. The large amount of earth so produced is well seen in the Gwalior State, where the late ruler employed an irrigation officer, lent by the Government of India, to construct a number of embankments, each furnished with spillways, across many of the valleys, which had suffered so badly by uncontrolled rain wash in the past that they appeared to have no soil at all, the scrub vegetation just managing to survive in the crevices of the bare rock. How great is the annual formation of new soil, even in such unpromising circumstances, must be seen to be believed.

In a few years the construction of embankments was followed by stretches of fertile land which soon carried fine crops of wheat. A brief illustrated account of the work done by the late Maharaja of Gwalior would be of great value at the moment for introducing a much needed note of optimism in the consideration of this soil erosion problem. Things are not quite so hopeless as they are often made to appear.

Why is the forest such an effective agent in the prevention of soil erosion? The forest does two things: (1) the trees and undergrowth break up the rainfall into fine spray and the litter on the ground further protects the soil from the impact of the descending water stream; (2) the residues of the trees and animal life met with in all woodlands are converted into humus, which is then absorbed by the soil underneath, increasing its porosity and water-holding power; the soil cover and the soil humus together prevent erosion and at the same time store large volumes of water. These factors — soil protection, soil porosity, and water retention — conferred by the living forest cover, provide the key to the solution of the soil erosion problem. All other purely mechanical remedies, such as terracing and drainage, are secondary matters, although, of course, important in their proper place.

The secret of soil conservation is thus seen to lie, first, in maintaining the soil cover in good condition to ensure that the rainfall is received on the surface in a proper manner with no disturbance of the soil below, and second, in conserving ample supplies of humus so that by means of the compound soil particles the water, when it has descended, is adequately absorbed and stored: as well might we expect a living creature to survive without its protective skin as to suppose that the earth can live without her proper covering. The forest has been cited as the pre-eminent example of these protective devices, for the leafage is thick and the ground litter abundant. In the absence of forest some form of grass cover is the natural protective agent which will for centuries often maintain the soil in good heart. Indeed, this device of the grass cover is far more efficient than might be supposed possible.

The accumulations of humus under a grass carpet are often immense; they are, indeed, so extraordinary that they can be described as veritable mines of fertility. This is proved by the fact that an agriculture based on their spoliation can, in favourable circumstances, continue for many years before it fades out. But fade out it must if the humus is never restored. Williams regarded grass as the basis of all agricultural land utilization and the soil's chief weapon against the plundering instincts of humanity. He advanced the hypothesis that the decay of past civilizations was due to the wholesale ploughing up of grass necessitated by the increasing demands of civilization. His views are exerting a marked influence on soil conservation policy in the U.S.S.R. and indeed apply to many other countries.

Grass is a valuable factor in the correct design and construction of surface drains. Whenever possible these should be wide, very shallow, and completely grassed over. The runoff then drains away as a thin sheet of clear water, leaving all the soil particles behind. The grass is thereby automatically manured and yields abundant fodder. This simple device was put into practice at the Shahjahanpur Sugar Experiment Station in India. The earth service roads and paths were excavated so that the level was a few inches below that of the cultivated area. They were than grassed over, becoming very effective drains in the rainy season, carrying off the excess rainfall as clear water without any loss of soil.

If we regard erosion as the natural consequence of improper methods of agriculture and the catchment area of the river as the natural unit for the application of soil conservation methods, the various remedies available fall into their proper place. The upper reaches of each river system must be afforested; cover crops, including grass and leys, must be used to protect the arable surface whenever possible; the humus content of the soil must be increased and the crumb structure restored, so that each field can drink in its own rainfall; overstocking and over-grazing must be prevented; simple mechanical methods for conserving the soil and regulating the runoff, like terracing, contour cultivation and contour drains, must be utilized. There is, of course, no single anti-erosion device which can be universally adopted. The problem must, in the nature of things, be a local one. Nevertheless, certain guiding principles exist which apply everywhere. First and foremost is the restoration and maintenance of the crumb structure of the soil, so that each acre of the catchment area can do its duty by absorbing its share of the rainfall.

The Formation of Alkali Land

When the land is continuously deprived of oxygen, the plant is soon unable to make use of the nourishment it contains: it becomes a dead instrument, from which no crop can draw anything. If left to itself, this condition of infertility is permanent.

In many parts of the tropics and sub-tropics agriculture is interfered with and even brought to an end because of the injury inflicted on the soil by accumulations of soluble salts composed of various mixtures of the sulphate, chloride, and carbonate of sodium. Such areas are known as alkali lands. When the alkali phase is still in the mild or incipient stage, crop production becomes difficult and care has to be taken to prevent matters from getting worse. When the condition is fully established, the soil dies; crop production is then out of the question. Alkali lands are common in Central Asia, India, Persia, Iraq, Egypt, North Africa, and the United States.

At one period it was supposed that alkali salts were the natural consequences of a light rainfall, insufficient to wash out of the land the salts which always form in it by progressive weathering of the rock powder, of which all soils largely consist. Hence alkali lands were considered to be a natural feature of arid tracts such as parts of north-west India, Iraq, and northern Africa, where the rainfall is very small. Such ideas of the origin and occurrence of alkali lands do not correspond with the facts and are quite misleading. The rainfall of the Province of Oudh in India, for example, where large stretches of alkali lands naturally occur, is certainly adequate to dissolve the comparatively small quantities of soluble salts found in these infertile areas, if their removal were a question of sufficient water only. In North Bihar the average rainfall in the submontane tracts where large alkali patches are common is about fifty to sixty inches a year. Arid conditions, therefore, are not essential for the production of alkali soils; heavy rainfall does not always remove them.

What is a necessary condition is impermeability. In India, whenever the land loses its porosity by the constant surface irrigation of stiff soils with a tendency to impermeability, by the accumulation of stagnant subsoil water, or through some interference with surface drainage, alkali salts sooner or later appear. Almost any agency, even over-cultivation or over-stimulation by means of artificial manures, both of which oxidize the organic matter and slowly destroy the crumb structure, will produce alkali land. In the neighbourhood of Pusa in North Bihar old roads and the sites of bamboo clumps and of certain trees, such as the tamarind *(Tamarindus indica* L.) and the pipul *(Ficus religiosa* L.), always give rise to alkali patches when they are brought into cultivation.

The densely packed soil of such areas invariably shows the bluish-green markings which are associated with the activities of those soil organisms existing in badly aerated soils without a supply of free oxygen. A few inches below the alkali patches which occur on the stiff, loess soils of the Quetta valley, similar bluish-green and brown markings always occur. In the alkali zone in North Bihar wells have always to be left open to the air, otherwise the water is contaminated by sulphuretted hydrogen, thereby indicating a well-marked, reductive phase in the deeper layers.

In a subsoil drainage experiment on the black soils of the Nira valley in Bombay, where perennial irrigation was followed by the formation of alkali land, Mann and Tamhane found that the salt water which ran out of these drains soon smelt strongly of sulphuretted hydrogen and a white deposit of sulphur was formed at the mouth of each drain, proving how strong were the reducing actions in this soil. Here the reductive phase in alkali formation was unconsciously demonstrated in an area where alkali salts were unknown until the land was waterlogged by over-irrigation and the oxygen supply of the soil was restricted.

The view that the origin of alkali land is bound up with defective soil aeration is supported by the recent work on the origin of salt water lakes in Siberia. In Lake Szira-Kul between Bateni and the mountain range of Kizill Kaya, Ossendowski observed in the black ooze taken from the bottom of the lake and in the water a certain distance from the surface an immense network of colonies of sulphur bacilli, which gave off large quantities of sulphuretted hydrogen and so destroyed practically all the fish in this lake. The great water basins in central Asia are being metamorphosed in a similar way into useless reservoirs of salt water, smelling strongly of hydrogen sulphide. In the limans near Odessa and in portions of the Black Sea a similar process is taking place. The fish, sensing the change, are slowly leaving this sea as the layers of water, poisoned by sulphuretted hydrogen, are gradually rising towards the surface. The death of the lakes scattered over the immense plains of Asia and the destruction of the impermeable soils of this continent from alkali salt formation are both due to the same primary cause — intense oxygen starvation. In the instances just mentioned this oxygen starvation occurs naturally; in other cases it follows perennial irrigation.

Every possible gradation in alkali land is met with. Minute quantities of alkali salts in the soil have no injurious effect on crops or on the soil organisms. It is only when the proportion increases beyond a certain limit that they first interfere with growth and finally prevent it altogether. Leguminous crops are particularly sensitive to alkali, especially when this contains carbonate of soda. The action of alkali salts on the plant is a physical one and depends on the osmotic pressure of solutions, which increases with the amount of the dissolved substance. For water to pass readily from the soil into the roots of plants, the osmotic pressure of the cells of the root must be considerably greater than that of the soil solution outside. When the soil solution becomes stronger than that of the cells, water passes backwards from the roots to the soil and the crops dry up. This state of affairs inevitably occurs when the soil becomes charged with alkali salts beyond a certain point. The crops are then unable to take up water and death results. The roots behave like a plump strawberry when placed in a strong solution of sugar; like the strawberry they shrink in size because they have lost water to the stronger solution outside. Too much salt in the water, therefore, makes irrigation water useless and destroys the canal as a commercial proposition.

The reaction of the crop to the first stages in alkali production is interesting. For twenty years at Pusa and eight years in the Quetta valley I had to farm land, some of which hovered, as it were, on the verge of alkali. The first indication of the condition is a darkening of the foliage and the slowing down of growth. Attention to soil aeration, to the supply of organic matter, and to the use of deep-rooting crops like lucerne and the pigeon pea, which break up the subsoil, soon set matters right. Disregard of Nature's danger signals, however, leads to trouble — a definite alkali patch is formed. When cotton is grown under canal irrigation on the alluvial soils of the Punjab, the reaction of the plant to incipient alkali is first shown by the failure to set seed, on account of the fact that the anther, the most sensitive portion of the flower, fails to function and to liberate its pollen. The cotton plant naturally finds it difficult to obtain from mild alkali soil all the water it needs — this shortage is instantly reflected in the breakdown of the floral mechanism.

Is the alkali condition confined to the tropics and sub-tropics? May it not, under certain circumstances, occur in temperate regions such as north-western Europe? Is it a factor in the sandy soils of Wareham in Dorsetshire recently investigated by Professor Neilson-Jones and Dr. Rayner? It is impossible at the moment to answer these questions till the soil studies of the future consider the biological activities in relation to the physical and chemical factors as well as to the season. They may not have reached the grade of decay known as alkali land, but they are starved of oxygen, all the conditions needed for the establishment of the anaerobic and semi-anaerobic state being present.

This is made clear by the readiness with which they respond to any improvement in surface and subsoil drainage, as well as to sub-soiling. Soil conditions must be looked at as a living and changing system and not merely as something static and stable. The soils of the north temperate zone, for example, often suffer from poor soil aeration. Moreover, many of the soil profiles exhibit the blue and red markings so common under alkali patches, as well as bands of humus which must have been originally formed near the surface, then carried in solution and afterwards precipitated. The soil organisms, which reduce compounds containing sulphur to sulphuretted hydrogen, are known to exist in these soils.

All facts point to the necessity for further work so as to provide a clear answer to the above mentioned questions, while from the practical point of view there is an immense field for improvement, especially by means of sub-soiling, over many areas which are now allowed to continue in a very unsatisfactory state. The problem of soil aeration is by no means, therefore, confined to the tropics, and it behoves the pioneers of farming in the temperate countries to turn an immediate attention to the various fairly simple devices by which very great, and above all, permanent improvements could be effected.

The stages in the development of the alkali condition are somewhat as follows. The first condition is an impermeable soil. Such soils —

the *usar* plains of northern India for example — occur naturally where the climatic condition favour those biological and physical factors which destroy the soil structure by disintegrating the compound particles into their ultimate units.

These latter are so extremely minute and so uniform in size that they form with water a mixture possessing some of the properties of colloids which, when dry, pack into a hard, dry mass, practically impenetrable to water and very difficult to break up. Such soils are very old. They have always been impermeable and have never come into cultivation.

In addition to the alkali tracts which occur naturally, a number are in course of formation as the result of errors in soil management, the chief of which are as follows:

(a) *The excessive use of irrigation water:* this gradually destroys the binding power of the organic cementing matter which glues the soil particles together, and further displaces the soil air. Anaerobic changes, indicated by blue and brownish markings, first occur in the lower layers and finally lead to the death of the soil. It is this slow destruction of the living soil that must be prevented if the existing schemes of perennial irrigation are to survive. The process is taking place before our eyes today in the Canal Colonies of India, where irrigation is loosely controlled.

(b) *Over-cultivation without due attention to the replenishment of humus:* in those continental areas like the Indo-Gangetic plain, where the risk of alkali is greatest, the normal soils contain only a small reserve of humus, because the biological processes which consume organic matter are very intense at certain seasons, due to sudden changes from low to very high temperatures and from intensely dry weather to periods of moist, tropical conditions. Accumulations of organic matter such as occur in temperate zones are impossible. There is, therefore, a very small margin of safety. The slightest errors in soil management will not only destroy the small reserve of humus in the soil, but also the organic cement on which the compound soil particles and the crumb structure depend. The result is impermeability, the first stage in the formation of alkali salts. The inhabitants of these areas through the centuries have followed methods of cultivation which are perfectly adapted to preserve the safety margin, but there is a tendency on the part of the shortsighted Western scientist to teach them so-called techniques of stimulating crop production which are highly dangerous from this point of view. One suggestion that is constantly being put forward is the introduction into the Indo-Gangetic plain of artificial manures like sulphate of ammonia. This would soon lead to catastrophe.

(c) *The use of artificial manures, particularly sulphate of ammonia:* even where there is a large safety margin, i.e. a large reserve of humus, such dressings do untold harm. The presence of additional combined nitrogen in an easily assimilable form stimulates the growth of fungi and other organisms which, in the search for the organic matter needed for energy and for building up microbial tissue, use up first the reserve of soil humus and then the more resistant organic matter which cements the soil particles. This glue is not affected by the processes going on in a normally cultivated soil, but it cannot withstand the same processes when stimulated by dressings of artificial manures.

Alkali land, therefore, starts with a soil in which the oxygen supply is permanently cut off. Matters then go from bad to worse very rapidly. All the oxidation factors which are essential for maintaining a healthy soil cease. A new soil flora — composed of anaerobic organisms which obtain their oxygen from the sub-stratum — is established. A reduction phase ensues. The easiest source of oxygen — the nitrates — is soon exhausted. The organic matter then undergoes anaerobic fermentation. Sulphuretted hydrogen is produced as the soil dies, just as in the lakes of central Asia.

The final result of the chemical changes that take place is the accumulation of the soluble salts of alkali land — the sulphate, chloride, and carbonate of sodium. When these salts are present in injurious amounts, they appear on the surface in the form of snow-white and brownish-black incrustations. The former (white alkali) consists largely of the sulphate and chloride of sodium, and the latter (the dreaded black alkali) contains sodium carbonate in addition and owes its dark colour to the fact that this salt is able to dissolve the organic matter in the soil and produce physical conditions which render drainage impossible. According to Hilgard, sodium carbonate is formed from the sulphate and chloride in the presence of carbon dioxide and water. The action is reversed in the presence of oxygen. Subsequent investigations have modified this view and have shown that the formation of sodium carbonate in soil takes place in stages. The appearance of this salt always marks the end of the chapter. The soil is dead. Reclamation then becomes difficult on account of the physical conditions set up by these alkali salts and the dissolved organic matter.

The occurrence of alkali land, as would be expected from its origin, is extremely irregular. When ordinary alluvial soils like those of the Punjab and Sind are brought under perennial irrigation, small patches of alkali first appear where the soil is heavy; on stiffer areas the patches are large and tend to run together. On open, permeable stretches, on the other hand, there is no alkali. In tracts like the western districts of the United Provinces, where irrigation has been the rule for a long period, zones of well aerated land carrying fine irrigated crops occur alongside the barren alkali tracts. Iraq also furnishes interesting examples of the connection between alkali and poor soil aeration. Intensive cultivation under irrigation is only met with in that country where the soils are permeable and the natural drainage is good. Where the drainage and aeration are poor the alkali condition at once becomes acute. There are, of course, a number of irrigation schemes, such as the staircase cultivation of the Hunzas in northwest India and of Peru, where the land has been continually watered from time immemorial without any development of alkali salts. In Italy and Switzerland perennial irrigation has been practiced for long periods without harm to the soil. In all such cases, however, careful attention has been paid to drainage and aeration and to the maintenance of humus; the soil processes have been confined by Nature or by man to the oxidative phase; the cement of the compound particles has been protected by keeping up a sufficiency of organic matter.

The theory of the reclamation of alkali land is very simple. All that is needed, after treating the soil with sufficient gypsum (which transforms the sodium clays into calcium clays), is to wash out the soluble salts, to add organic matter, and then to farm the land properly. Such reclaimed soils are then exceedingly fertile and remain so. If sufficient water is available, it is sometimes possible to reclaim alkali soils by washing only. I once confirmed this. The berm of a raised water channel at the Quetta Experiment Station was faced with rather heavy soil from an alkali patch. The constant passage of the irrigation water down the water channel soon removed the alkali salts. This soil then produced some of the heaviest crops of grass I have ever seen in the tropics. When, however, the attempt is made to reclaim alkali areas on a field scale by flooding and draining, difficulties at once arise unless steps are taken first to replace all the sodium in the soil complex by calcium and then to prevent the further formation of sodium clays. Even when these reclamation methods succeed, the cost is always considerable; it soon becomes prohibitive; the game is not worth the candle.

The removal of alkali salts is only the first step; large quantities of organic matter are then needed; adequate soil aeration must be provided; the greatest care must be taken to preserve these reclaimed soils and to see that no reversion to the alkali condition occurs. It is exceedingly easy under canal irrigation to create alkali salts on certain areas. It is exceedingly difficult to reverse the process and to transform alkali land back again into a fertile soil.

An interesting development in the reclamation of alkali soils has recently taken place at the Coleyana Estate in the Montgomery District of the Punjab. The method adopted is a first-rate pointer to the right way of solving this or any other agricultural problem. It consists in a clever diagnosis of natural processes and an ingenious adaptation of them to attain the wished-for end. Nature is made, as it were, to retrace certain steps so as to re-establish more desirable soil conditions; she is asked to undo her own work. On the Coleyana Estate Colonel Sir Edward Hearle Cole, C.B., C.M.G., first removes the accumulations of alkali salts from the surface, then ploughs them up and plants *dhup* grass *(Cynodon dactylon,* Pers.) which is grazed as heavily as possible by sheep and cattle for some eighteen months to two years.

The turf is then killed by a turnover plough followed by a fallow during the hot season (May and June). The land is then prepared for a green-manure crop, followed by a couple of wheat crops in succession, and then put into lucerne or cotton. The great thing in this reclamation work is to scrape off all alkali salts as they appear, remove them from the land, and use the minimum irrigation water for the establishment and maintenance of the crop of grass. The underground stems and roots of the grass then aerate the heavy soil: the sheet-composting of the turf and the droppings of the livestock create the large quantities of humus needed to get this heavy land into condition for wheat, cotton, and lucerne. Sir Edward is now making a point of never leaving such reclaimed land uncovered so as to make the fullest use of the energy of sunlight in creating vegetable matter, which ultimately gets converted into humus. He also takes advantage of deep-rooting plants such as chicory, lucerne, and *arhar (Cajanus indicus,* Spreng.) for breaking up the subsoil and is a firm believer in the principles set out in *The Clifton Park System of Farming.* In this way, areas once ruined by alkali salts are now producing crops of wheat up to 1,600 lb. to the acre. This is, perhaps, the simplest and easiest method of reclaiming alkali soils that has yet been devised. It makes the crop itself do most of the work.

A further development of the Coleyana method of reclaiming alkali land suggests itself. When the grass crop is ploughed up, it might be worth while to subsoil the land to a depth of fifteen to eighteen inches four feet apart, using a caterpillar tractor and a Ransomes sub-soiler. This would shatter the deeper soil layers, provide abundant aeration, and prepare the land for the succeeding crops.

Nature has provided, in the shape of alkali salts, a very effective censorship for all schemes of perennial irrigation. The conquest of the desert by the canal by no means depends on the mere provision of water and arrangements for the periodical flooding of the surface. This is only one of the factors of the problem. The water must be used in such a manner and the soil management must be such that the fertility of the soil is maintained intact. There is obviously no point in creating at vast expense a Canal Colony and producing crops for a generation or two, followed by a permanent desert of alkali land. Such an achievement merely provides another example of agricultural banditry.

It must always be remembered that the ancient irrigators never developed any efficient method of perennial irrigation, but were content with the basin system, a device by which irrigation and soil aeration can be combined. (The land is embanked; watered once; when dry enough it is cultivated and sown. In this way water can be provided without any interference with soil aeration.) In his studies on irrigation and drainage, King concludes an interesting discussion of this question in the following words which deserve the fullest consideration on the part of the irrigation authorities all over the world:

> *It is a noteworthy fact that the excessive development of alkalis in India, as well as in Egypt and California, is the result of irrigation practices modern in their origin and modes and instituted by people lacking in the traditions of the ancient irrigators, who had worked these same lands thousands of years before. The alkali*

*lands of today, in their intense form, are of modern
origin, due to practices which are evidently inadmissible,
and which in all probability were known to be so by the
people whom our modern civilization has supplanted.f*

These words should be studied by all who are concerned with the
extension of irrigation schemes. The unwise pursuance of such schemes
with a view to the immediate production of easily grown crops without
the lasting maintenance of fertility can only end in the regular
suffocation of precious tracts of the earth's surface.

The Maintenance of Soil Fertility in Great Britain

Many accounts of the way the present system of farming in Great
Britain has arisen have been published. The main facts in its evolution
from Saxon times to the present day are well known. Nevertheless,
in one important respect these surveys are incomplete. Nowhere has
any attempt been made to bring out the soil fertility aspect of this
history and to show what has happened all down the centuries to that
factor in crop production and animal husbandry — the humus content
of the soil — on which so much depends. The present chapter should
be regarded as an attempt to make good this omission.

The Roman Occupation

At the time of the Roman invasion most of the island in which
we are living was under forest or marsh: only a portion of the uplands
was under grass or crops: the population was very small. After the
conquest of the country the Romans began to develop it by the creation
on the areas already cleared of an agricultural unit — new to Great
Britain — known as the villa. These villas were large farms under
single ownership run by functionaries each responsible for a particular
type of animal or crop and worked by slave labour. These units
followed to some extent the methods of the *latifundia* of Italy and
were designed for the production of food for the legions garrisoning
the island and those stationed in Gaul. Wheat — an exhausting crop
— was an important item in Roman agriculture, for the reason that
this cereal provided the chief food *(frumentum)* of the soldiers. The
extent of the export of grain to Gaul will be evident from the fact that
in the reign of the Emperor Julian no less than 800 wheat ships were
sent from Britain to the Continent.

The exhaustion of the soils of the island began even before the
Roman occupation. The heavy soil-inverting mould board plough,
which invariably wears out the land, was already in use when the

Romans arrived, and was probably brought by the Belgic tribes who conquered and settled in the south-eastern part of the country. They lived in farmsteads and cultivated large open fields. They were highly skilled agriculturists and exported to Gaul a considerable quantity of their main product — wheat. This practice was developed by the Roman villas which followed and in this way the slow exhaustion of the lighter soils of the downlands of the southeast became inevitable.

After an occupation which lasted some 400 years and which contributed little or nothing of permanent value to the agriculture of the island beyond some well-designed roads, the legions evacuated the island and left the Romanized population to look after itself. This they failed to do: the country was soon conquered by the Saxon invaders, in the course of which much destruction of life and property took place. One result was the creation of a new type of farming.

The Saxon Conquest

The settlement of Nordic people in our island is the governing event both of British history and of British agriculture. The new settlers had inhabited the belts of land around the Weser and the Elbe and their first contact with Britain was as raiders; their operations were in the nature of reconnaissance to ascertain the chances of settlement. The Anglo-Saxon migration to Britain was a colonization preceded by conquest, in which the farming system of the Romanized population was, in the midland area at any rate, destroyed. In the east, southeast, and western portions of the island some relics of Roman and Celtic methods survived. Our forefathers brought with them from the opposite shores of the North Sea their wives, children, livestock, and a complete fabric of village life. The immigrants, being country folk, wanted to live in rural huts with their cattle round them and their land nearby, as they did in Germany. The numerous villages they formed reproduced in all essentials those they had left behind on the mainland. Our true English villages are, therefore, not Celtic, are not Roman, but purely and typically German.

The Roman villas were replaced by a new system of farming — the Saxon manor — in which the tenants held land in return for service. The lord and his retainers shared the land, each bound to perform certain duties determined by custom. The manors took centuries to evolve. By A.D. 800 they had developed into a permanent system which provided the material for the Domesday Book of the Normans, by which taxation was assessed and a rigid feudal system became firmly established.

The Open-Field System

The first general feature that strikes us in early Anglo-Saxon England is the strip cultivation of the arable land on the open-field system. This system was a communal agricultural institution started by people who had to get a living out of the soil. They had progressed as far as to use the plough and had a common fund of experience. Everyone pursued the same system of farming. The arrangement of the open fields was, however, by no means uniform. No fewer than three distinct types arose, corresponding to as many different influences exerted by people who had early occupied the country. The large central midland area, stretching from Durham to the Channel and from Cambridgeshire to Wales, is the region where Germanic usage prevailed.

The southeast was characterized by the persistence of Roman influence, a circumstance which implies that the conquest was less destructive there than in the north and west. The counties of the south-west, northwest, and the north retained Celtic agrarian usages in one form or another, which is easily understood in view of the difficulty with which, as we know, these districts were slowly overpowered by the invaders. The midland area was thus the region where the Anglo-Saxons were most firmly established and where the subjugation of the fifth century was most thorough. The Romano-Celtic people who remained were not numerous enough to preserve any traces of Roman or Celtic methods of tilling the soil.

Throughout this extensive region a two-field and a three-field system, or sometimes a mixture of the two, prevailed. This field arrangement was a custom prevalent in Germany, especially east and south of the Weser. The chief characteristic of the two- and three-field type of tillage was the distribution of the parcels of arable land (which made up the holdings of the customary tenants) equally amongst the two or three fields. The cropping was so arranged that one field in the two-field system and two fields in the three-field system were cropped every year, and thus one-half or one-third of the township's arable land lay fallow and was used for common grazing — a point which is always emphasized in the midland system.

Besides the cultivated open fields, for which the best land was always used, the village lands consisted of grassland for mowing on the wetter parts, and commons or woodlands on the poorer parts.

Ploughing was the all-important operation of medieval tillage and was carried out on a co-operative basis, and demanded a team of eight draught animals yoked to a heavy plough. This, of course, was beyond the reach of any but the largest and most prosperous tenants. Communal ploughing in Saxon times was, therefore, inevitable. It was the difficulty of replacing this communal ploughing that delayed agricultural progress in many parts of the country.

The open-field system repeated itself for centuries, not only in England but in a great part of Europe — nations living under very different conditions, in very different climates, and on very different soils adopted the open-field system again and again without having borrowed it from each other. This could not but proceed from some pressing necessity. The open-field system is communal in its very essence. Every trait which makes it strange and inconvenient from the point of view of individualistic interests renders it highly appropriate to a state of things ruled by communal conceptions — right of common usage — communal arrangements of ways and time of cultivation. These are the main features of open-field husbandry and all point to one origin — the formation in early Anglo-Saxon society of a village community of shareholders of free and independent growth.

It must be borne in mind that the open-field prevailed during the period of national formation of the English people and its influence on the life of the village community must have been very great. The sense of personal responsibility, which the system of communal work created, made it a vital factor in the social education of the people.

The Depreciation of Soil Fertility

Open-field farming is, as a rule, balanced: the fertility used up in growth is made good before the next crop is sown. Compared with our modern standards, however, the yield is remarkably low and the removal of fertility by such small crops is made up for by the recuperative processes operating in the soil (non-symbiotic fixation of nitrogen and so forth). The surplus of available humus originally left by the forest is depleted at an early stage and an equilibrium is established, the yield adjusting itself to the amount of fertility added each year by natural processes, this in its turn is influenced by climate and methods of cultivation.

For example, in the peasant cultivation of northwest India at the present day a perfect balance has been established between losses and gains of fertility. The village land on which corn crops are grown has been cultivated for upwards of 2,000 years without manure beyond the droppings of the livestock during the fallow period between harvest and the rains. But the Indian cultivators use primitive scratch ploughs and are most careful not to draw on the reserves of organic material in the soil, as its texture depends on this. They produce crops entirely on the current account provided by the annual increments of fertility. The yield has settled down to 8 maunds (658 lb. per acre) of wheat on unirrigated land, and 12 maunds (987 lb.) of wheat on irrigated land, and this yield has been constant for many centuries.

The same processes were operating in the English open fields. The reserve of humus in the soils originally under forest, which the Saxons brought into cultivation, was soon used up and the yield was determined by the annual additions of fertility to the soil by natural means. But in our cold and sunless climate and on our ill-drained, poorly aerated soils this is far less than in the semi-tropical conditions of northern India.

Moreover, and this point must be stressed, the Saxons from the earliest times used a soil-inverting plough, which has a marked tendency to exhaust the humus in the soil if provision is not made for the regular supply of sufficient farmyard manure. In fact, recent experience in many parts of the world is proving that the continued use of heavy soil-inverting, tractor-driven implements, without sufficient farmyard manure to manure the land, promptly leads to catastrophic consequences.

The first recorded references to the mould board plough speak of it in Gaul, but some authorities quoted by Vinogradoff (*The Growth of the Manor*) suggest that it was borrowed by the Germanic people from the Slavs, and in view of the soil types found in Slav territory this may easily be so. The evolution of the big plough was due to soil requirements as settled agricultural life developed in the heavy, moist soils of north Europe after the forests had been cleared.

The mould board plough determined the lay-out of the open fields. It divided the arable areas into a succession of lands. It needed a headland to turn on, and there was a limit to the length of furrow a team of oxen could plough before needing the relief got by stopping and turning. This furrow- long or furlong became one of our units of length. It was usual to keep the land in high ridges running along the slopes to facilitate surface drainage, an important point in England. The ridges varied in width according to the nature of the soil. In very heavy clays they were sometimes no more than three yards wide. In lighter soils they might be twenty-two yards wide. These ridges may be seen in many places today on grassland which was under the plough in earlier centuries. From this brief description it will be seen that the open fields cultivated with the heavy medieval plough were laid out in strips.

The main feature of the heavy mould board plough was its high penetrating power, and it could be used on the heavier types of soil where the light scratch plough of the Celts and Italians would be useless. It thus enabled the cropped area in England to be greatly extended by the cultivation of the heavy soil of the valleys and plains which first had to be slowly carved out of the forest. It owed its superiority to an iron share, a courter, and a wooden mould board so suitable on wet land. This primitive implement gave us the plough as we know it today. The principle of our modern plough is identical and, except for the fact that it is now made entirely of iron, it is almost the same in detail.

The open-field system of the Middle Ages was bound to fail because it involved burning the candle at both ends and also in the middle. First the natural recuperation processes in the soil were hampered by low temperatures and poor soil aeration; second, such supplies of farmyard manure as were available were by custom mostly bestowed on the lord's demesne lands, and besides were inadequate because only a portion of the livestock could be wintered; finally the soil-inverting plough led to the oxidation of the stores of soil humus faster than it could be recreated and was bound to wear out the land.

The Low Yield of Wheat

The failure of the open-field system is proved by the low yield of wheat. All authorities agree that the yield of wheat in England during the Middle Ages was at a very low level, though it does not appear to have varied greatly. It may be noted that there was never any question of complete exhaustion of the wheat-growing land, such as occurred in Mesopotamia and in the Roman wheat-growing regions of North Africa, where the soil, owing to over-cropping and in some instances to over-irrigation aggravated by special climatic conditions, became sterile and was transformed into desert.

This could not so easily happen in the moist, temperate climate of Great Britain. What happened in the Middle Ages in England was that the yield of corn was not high enough for the requirements of the growing social and economic life of the country.

The material for a quantitative estimate of wheat yields in this period is necessarily very scanty, but in the case of some large estates records are available for a considerable period of years of the seed sown in one year and the grain threshed in the following year, and these form the basis of the best estimates of medieval yields. Sir William Beveridge, using this method, investigated the yield of wheat for the years 1200 to 1450 on eight manors, including that of Wargrave, situated in seven different counties belonging to the Bishop of Winchester. The average yield per acre was 1.17 quarters or 9.36 measured bushels, equivalent to 7.48 bushels of 60 lb. It is to be noted that these estimates were all from demesne lands which were probably better cultivated and better manured than the land of the customary tenants. Other authorities confirm these figures.

The figures of yield given above help to account for the changes which marked the end of the Middle Ages. The amount of food was becoming insufficient for the growing population. But another factor was steadily developing, which finally assumed the dimensions of an avalanche and led to the reform of manorial farming. This was disease, a matter which must now be discussed.

The Black Death

That the agriculture of the Middle Ages was unable to keep the population in health was first indicated by the frequent indications of rural unrest. But these were soon followed by the writing on the wall in the shape of the Black Death in 1348-9. This outbreak had been preceded by several years of dearth and pestilence, and it was succeeded by four visitations of similar disease before the end of the century. During its ravages it destroyed from one-third to one-half of the population. This seriously affected the labour supply, which was no longer sufficient to carry on the traditional methods of manorial farming, already beginning to be undermined by the growing tendency to replace service by money payments.

Land which could no longer be ploughed had to be laid down to grass and used for feeding sheep to produce more of the wool so urgently needed in Flanders and Lombardy. For the new farming the countryside had to be enclosed: first the lord's demesne and then the area under open fields began to be laid down to grass. The earth's green carpet not only fed the sheep, but gave the land a long rest: large reserves of humus were gradually built up under the turf: the fertility of the soil, which had been imperceptibly worn out by the mould board plough and the constant cropping of the manorial system, was gradually restored.

After a long period of rest of a century the land no longer returned only seven and a half bushels to the acre. The figures given above for the years 1200 to 1450 may be contrasted with the figures from a farm at Wargrave from 1612-20: in these years the average was 25.6 bushels of 60 lb. per acre. In the latter part of the sixteenth century the general average was eighteen bushels to the acre and even more. That this significant change was due to the restoration of soil fertility by humus formation under the turf there can be no doubt.

It is more than probable that the slow regeneration of the soils of this country, which began after the Black Death, produced other results besides the improvement of crops and livestock. What of the effect of the produce of land in good heart on the most important crop of all — men and women? Were the outstanding achievements of the Tudor period one of the natural consequences of a restored agriculture? It may well be so.

Enclosure

When increasing population led once more to the breaking up of the grassland and the farmer returned to tillage, the land, after its long rest of upwards of a century, was again capable of responding to the demands made upon it. One result of this experience was an increased interest in enclosure. Instinct was leading to a search for an economic arrangement which would prevent soil exhaustion from being repeated in succeeding ages. Enclosed farms offered a solution, as they gave the farmer the chance of keeping his land in good condition by individual management in place of the easy-going farming of the open fields of old English village agriculture. They also offered to the enclosed farmer the opportunity of composting his straw in his cattle yards and producing as much farmyard manure as possible. This, in most cases, he did, and the plan succeeded.

Nevertheless, the ancient open-field tillage husbandry had in its favour the authority of long tradition — a potent force with a suspicious and conservative peasantry. The peasant asked himself: In the case of a readjustment of holdings would not the strong profit and the weak suffer? There grew up a popular prejudice against enclosure and the improvement of the common fields, but in the end, after some centuries of contest, enclosure won.

The form which the enclosure movement took before it was completed was due to the peculiar form of government which came in with the English Revolution of 1688. By that event the landed gentry became supreme. The national and local administration was entirely in their hands, and land, being the foundation of social and political influence, was eagerly sought by them. They not unnaturally wished to direct the enclosure movement into channels which were in the interests of their estates. But in doing so they made some of the most outstanding contributions to farming ever made in our history.

The restoration of soil fertility which resulted from enclosure had a profound influence on both livestock and crops. The provision of more and better forage and fodder which followed the cultivation of clover and artificial grasses, coupled with the popularization of the turnip crop by Townshend in 1730, opened the door for the continuous improvement of livestock by pioneers like Bakewell. The result was that our livestock improved in size and in the quality of the meat. Between 1710 and 1795 the weights of cattle sold at Smithfield more than doubled. By 1795 beeves weighed 800 lb. as compared with 370 lb.; sheep went up from 28 lb. to 80 lb. The improvement in the yield of cereals was no less significant. That of rye or wheat rose from 6-8 bushels to the acre in the Middle Ages to 15-20 bushels; barley yielded up to 36 bushels, oats 32-40 bushels. All this was due to more and better food for the livestock and more manure for the land. More manure raised larger crops: larger crops supported much bigger flocks and herds.

Another change in the countryside accompanied the enclosures. The forests, which since Saxon times had been gradually cleared and converted into manorial lands, had by this process become exhausted. After the Civil War it was realized that the country was running short of the hardwoods needed for maintaining the fleet and for buildings and so forth. An era of tree planting, which continued for two hundred years, was inaugurated by the publication of Evelyn's *Sylva* in 1678. It was during this period that the English landscape as we know it today was created by the judicious laying out of parks, artificial lakes, groups of trees, and woods.

All this planting provided an important factor in the maintenance of soil fertility. The roots of the trees and the hedges combed the subsoil for minerals, embodied these in the fallen leaves and other wastes of the trees and shrubs, and so helped to maintain the humus in the soil, as well as the circulation of minerals. The roots also acted as subsoil ploughs and aerating agencies. The cumulative effect of the trees and hedges, which accompanied enclosure, in maintaining soil fertility has passed almost unnoticed. Nevertheless, its importance in humus production and in the availability of minerals must be considerable.

While the policy of enclosure, combined with tree-planting and the creation of the existing English landscape, arrested the fall in soil fertility which was inherent in the open-field system, the freedom of action which followed enclosure afforded full scope to the improver. The restoration of British agriculture owes much to the pioneers among the landlords themselves, particularly to Coke of Holkham (1776-1816), who did much to introduce the Norfolk four-course system— (1) turnips, (2) barley, (3) seeds (clover and rye grass), (4) wheat — into general practice and so to achieve at long last an approach to Nature's law of return. Besides his championship of the Norfolk four-course system, his achievements include the conversion of 2,000,000 acres of waste into well-farmed and productive land, the prevention of famine in England during the Napoleonic Wars, the solution of the rural labour problem in his locality by means of a fertile soil, the demonstration of the principle that money well laid out in land improvement is an excellent investment. He invested half a million sterling in his own property and thereby raised the rent roll of his estate from £2,200 a year to £20,000. He transformed agriculture in this country by the simple process of first writing his message on the land and then, by means of his famous sheep-shearing meetings, bringing it to the notice of the farming community.

But the replacement of the manorial system by individual farming in fenced fields was attended by some grave disadvantages. The large profits obtained from the sale of wool, for example, while they enriched the few, led to a new conception of agriculture. The profit motive began to rule the farmer; farming ceased to be a way of life and soon became a means of enrichment. Enterprising individuals were afforded considerable scope for using their farms to make money. At the same time, large numbers of less fortunate individuals deprived of their land had either to work for wages or seek a living in the towns.

The various Enclosure Acts, which covered a period of more than 600 years, 1235-1845, therefore led to a new agriculture, the enthronement of the profit motive in the national life, and to the exploitation of coal, iron, and minerals, which is customarily referred to as the Industrial Revolution. This arose from the activities of the tradesmen of the manor, whose calling was destroyed by the Enclosure Acts.

The last of the Enclosure Acts, which finally put an end to the strip system of the open fields, was passed in 1845. About the same time the celebrated Broadbalk wheat plots of the Rothamsted Experimental Station were laid out. This field is divided into permanent parallel strips and cultivated on even more rigid lines than anything to be found in the annals of manorial farming. These plots never enjoy the droppings of livestock: till recently they never had the benefit of the annual rest provided by a fallow. Practically every agricultural experiment station all over the world has copied Rothamsted and adopted the strip system of cultivation. How can such experiments, based on an obsolete method of farming, ever hope to give a safe lead to practice? How can the higher mathematics and the ablest statistician overcome such a fundamental blunder in the original planning of these trials?

The strip system has also been adopted for the allotments round our towns and cities without any provision whatsoever on the part of the authorities to maintain the land in good heart by such obvious and simple expedients as subsoiling, followed by a rest under grass grazed by sheep or cattle, ploughing up, and sheet-composting the vegetable residues. Land under allotments should not be under vegetables for more than five years at a time; this should be followed by a similar period under grass and livestock.

The Industrial Revolution and Soil Fertility

The released initiative which accompanied the collapse of the manorial system was by no means confined to the restoration of soil fertility and the development of the countryside. The dispossessed craftsmen started all kinds of industries, in which they used as labour-saving devices first water power, then the steam engine, the internal combustion engine, and finally electrical energy. By these agencies the Industrial Revolution, which continues till this day, was set in motion. It has influenced farming in many directions. In the first place, industries have encroached on and seriously reduced the area under cultivation. But by far the most important demand of the Industrial Revolution was the creation of two new hungers — the hunger of a rapidly increasing urban population and the hunger of its machines. Both needed the things raised on the land: both have seriously depleted the reserves of fertility in our soils.

Neither of these hungers has been accompanied by the return of the respective wastes to the land. Instead, vast sums of money were spent in completely sidetracking these wastes and preventing their return to the land which so sadly needed them. Much ingenuity was devoted to developing an effective method of removing the human wastes to the rivers and seas. These finally took the shape of our present-day water-borne sewage system. The contents of the dustbins of house and factory first found their way into huge dumps and then into incinerators or into refuse tips sealed by a thin covering of cinders or soil.

At first the additional demands for food and raw materials were met by the restored agriculture and the periodical ploughing up of grass. One of these demands was the vast quantities of corn needed to feed the urban population. The price of wheat was regulated for more than 150 years by a series of Corn Laws, which attempted to hold the balance between the claims of the farmers who produced the grain and those of the consumers and the industrialists who advocated cheap food for their workers, so that they could export their produce at a profit. But as the urban population expanded, the pressure on the fertility of the soil increased until, in 1845, a disastrous harvest and the potato famine compelled the Government in 1846 to yield. The "rain rained away" the Corn Laws (Prothero).

Deprived of protection, farmers were forced to adopt new methods and to farm intensively. Many developments in farming occurred. Particular attention was paid to drainage: the first drain pipe was made in 1843; two years later the pipes were turned out by a machine. Liebig's famous essay in 1843 drew attention to the importance of manures, While better farm buildings and the preparation of better farmyard manure were adopted, two fatal mistakes were made. Artificial manures like nitrate of soda and superphosphate came into use: imported feeding stuffs for livestock began to take the place of home-grown food. British farming, in adopting these two expedients, because they appeared for the moment to be profitable, laid the foundations of much future trouble But in the use of better implements for the land and the provision of improved transport facilities the countryside was on firmer ground. The result of all these and other developments was a period of great prosperity for farming which lasted till late in the seventies of the last century.

The Great Depression of 1879

Then the blow fell. The year 1879, which I remember so vividly, was one of the wettest and coldest on record. The average yield of wheat fell to about fifteen bushels to the acre: large numbers of sheep and cattle were destroyed by disease: the price of wheat fell to an undreamt-of level as the result of large importations from the virgin lands of the New World. The great depression of 1879 not only ruined many farmers, but it dealt the industry a mortal blow. Farmers were compelled to meet a new set of conditions — impossible from the point of view of the maintenance of soil fertility — which have been more or less the rule till the Great War of 1914-18 and the World War which began in 1939 provided a temporary alleviation as far as the sale of produce and satisfactory prices were concerned.

Since 1879 the standard of real farming in this country has steadily fallen. The labour force, particularly the supply of men with experience of and sympathy with livestock, markedly diminished and deteriorated in quality. Rural housing left much to be desired. Drainage was sadly neglected. The small hill farms, which are essential for producing cattle possessing real bone and stamina, fell on evil days. Our flocks of folded sheep, so essential for the upkeep of downland, dwindled. Diseases like foot-and-mouth, tuberculosis, mastitis, and contagious abortion became rampant. Less and less attention was paid to the care of the manure heap and to the maintenance of the humus content of the soil. The Shortcomings of Present-day Agricultural Research) replaced the muck mentality of our fathers and grandfathers. Murdered bread, deprived of the essential germ, replaced the real bread of the last century and seriously lowered the efficiency of our rural population. The general wellbeing of our flocks and herds fell far below that of some of our overseas competitors like the Argentine.

But in this dark picture some rays of light could be detected. The pioneers were busy demonstrating important advances. Among these two are outstanding: (1) the Clifton Park system of farming based on deep-rooting plants in the grass carpet, and (2) the use of the subsoiler for breaking up pans under arable and grass, and so preparing the ground for another great advance — the mechanized organic farming of tomorrow.

The Second World War

Such, generally speaking, was the condition of British agriculture in September 1939, when the second world war began and the submarine menace for the second time brought national starvation into the picture. What an opportunity was provided for a Coke of Norfolk for making use of a portion of the resources of a great nation to set British farming on its feet for all time by the simple expedient of restoring and maintaining soil fertility! What an opening was given to the pioneers of human nutrition and the apostles of preventive medicine for feeding the men and women defending the country on the fresh produce of fertile soil and so initiating the greatest food reform in our history! But the potential Cokes of Norfolk had been liquidated or discouraged by many years of death duties, which had destroyed most of our agricultural capital and deprived the countryside of its natural leaders who, in years gone by, had done so much for farming. The apostles of real nutrition and of preventive medicine, such as the panel doctors of Cheshire, were ignored.

A much easier road was taken. The vast stores of fertility, which had accumulated after the long rest under grass, were cashed in and converted into corn crops. The seed so obtained saved the population from starvation, but most of the resulting straw could not be used because of the shortage of labour to handle it and of insufficient cattle to convert it into humus. The grow-more-food policy was, therefore, based on the exhaustion of the soil's capital. It is a perfect example of unbalanced farming. It is therefore certain to sow the seeds of future trouble, which will be duly registered by Mother Earth in the form of malnutrition and disease of crops, livestock, and mankind.

Chapter 7

Examination and Description of Soils

Introduction

A description of the soils is essential in any soil survey. This chapter provides standards and guidelines for describing most soil properties and for describing the necessary related facts. For some soils, standard terms are not adequate and must be supplemented by a narrative. The length of time that cracks remain open, the patterns of soil temperature and moisture, and the variations in size, shape, and hardness of clods in the surface layer must be observed over time and summarized.

This chapter does not include a discussion of every possible soil property. For some soils, other properties need to be described. Good judgment will decide what properties merit attention in detail for any given pedon (sampling unit). Observations must not be limited by preconceived ideas about what is important.

Although the format of the description and the order in which individual properties are described are less important than the content of the description, a standard format has distinct advantages. The reader can find information more rapidly, and the writer is less likely to omit important features. Furthermore, a standard format makes it easier to code data for automatic processing. If forms are used, they must include space for all possible information.

Each investigation of the internal properties of a soil is made on a soil body of some dimensions. The body may be larger than a pedon or represent a portion of a pedon. During field operations, many soils are investigated by examining the soil material removed by a sampling tube or an auger. For rapid investigations of thin soils, a small pit can be dug and a section of soil removed with a spade. All of these are samples of pedons. Knowledge of the internal properties of a soil is derived mainly from studies of such samples. They can be studied more rapidly than entire pedons; consequently, a much larger number can be studied in many more places. For many soils, the information obtained from such a small sample describes the pedon from which it is taken with few omissions. For other soils, however, important properties of a pedon are not observable in the smaller sample, and detailed studies of entire pedons may be needed. Complete study of an entire pedon requires the exposure of a vertical section and the removal of horizontal sections layer by layer. Horizons are studied in both horizontal and vertical dimensions.

Some General Terms Used in Describing Soils

Several of the general terms for internal elements of the soil are described here; other more specific terms are described or defined in the following sections.

A *soil profile* is exposed by a vertical cut through the soil. It is commonly conceived as a plane at right angles to the surface. In practice, a description of a soil profile includes soil properties that can be determined only by inspecting volumes of soil. A description of a pedon is commonly based on examination of a profile, and the properties of the pedon are projected from the properties of the profile. The width of a profile ranges from a few decimeters to several meters or more. It should be sufficient to include the largest structural units.

A *soil horizon* is a *layer*, approximately parallel to the surface of the soil, distinguishable from adjacent layers by a distinctive set of properties produced by the soil-forming processes. The term layer, rather than horizon, is used if all of the properties are believed to be inherited from the parent material or no judgment is made as to whether the layer is genetic.

The *solum* of a soil consists of a set of horizons that are related through the same cycle of pedogenic processes. In terms of soil horizons described in this chapter, a solum consists of A, E, and B horizons and their transitional horizons and some O horizons. Included are horizons with an accumulation of carbonates or more soluble salts if they are either within, or contiguous, to other genetic horizons and are judged to be at least partly produced in the same period of soil formation. The solum of a soil presently at the surface, for example, includes all horizons now forming. It includes a bisequum (to be discussed). It does not include a buried soil or a layer unless it has acquired some of its properties by currently active soil-forming processes. The solum of a soil is not necessarily confined to the zone of major biological activity. Its genetic horizons may be expressed faintly to prominently. A solum does not have a maximum or a minimum thickness.

Solum and soils are not synonymous. Some *soils* include layers that are not affected by soil formation. These layers are not part of the solum. The number of genetic horizons ranges from one to many. An A horizon that is 10 cm thick overlying bedrock is by itself the solum. A soil that consists only of recently deposited alluvium or recently exposed soft sediment does not have a solum.

In terms of soil horizons described in this chapter, a solum consists of A, E, and B horizons and their transitional horizons and some O horizons. Included are horizons with an accumulation of carbonates or more soluble salts if they are either within, or contiguous, to other genetic horizons and are judged to be at least partly produced in the same period of soil formation.

The lower limit, in a general sense, in many soils should be related to the depth of rooting to be expected for perennial plants assuming that water state and chemistry are not limiting. In some soils the lower limit of the solum can be set only arbitrarily and needs to be defined in relation to the particular soil. For example, horizons of carbonate accumulation are easily visualized as part of the solum in many soils in arid and semiarid environments. To conceive of hardened carbonate accumulations extending for 5 meters or more below the B horizon as part of the solum is more difficult. Gleyed soil material begins in some soils a few centimetres below the surface and continues practically unchanged to a depth of many meters. Gleying immediately below the A horizon is likely to be related to the processes of soil formation in the modern soil. At great depth, gleying is likely to be relict or related to processes that are more geological than pedological. Much the same kind of problem exists in some deeply weathered soils in which the deepest material penetrated by roots is very similar to the weathered material at much greater depth.

For some soils, digging deep enough to reveal all of the relationships between soils and plants is not practical. Roots of plants, for example, may derive much of their moisture from fractured bedrock close to the surface. Descriptions should indicate the nature of the soil-rock contact and as much as can be determined about the upper part of the underlying rock.

A *sequum* is a B horizon together with any overlying eluvial horizons. A single sequum is considered to be the product of a specific combination of soil-forming processes.

Most soils have a single sequum, but some have two or more. A Spodosol, for example, can form in the upper part of an Alfisol, producing an eluviated zone and a spodic horizon underlain by another eluviated zone overlying an argillic horizon. Such a soil has two sequa. Soils in which two sequa have formed, one above the other in the same deposit, are said to be bisequal.

If two sequa formed in different deposits at different times, the soil is not bisequal. For example, a soil having an A-E-B horizon sequence may form in material that was deposited over another soil that already had an A-E-B horizon sequence. Each set of A-E-B horizons is a sequum but the combination is not a bisequum. The lower set is a buried soil. If the horizons of the upper sequum extend into the underlying sequum, the affected layer is considered part of the upper sequum. For example, the A horizon of the lower soil may retain some of its original characteristics and also have some characteristics of the overlying soil. Here, too, the soils are not considered bisequal; the upper part of the lower soil is the parent material of the lower part of the currently forming soil. In many soils the distinction cannot be made with certainty. Nevertheless, the distinction is useful when it can be made. Where some of the C material of the upper sequum remains, the distinction is clear.

Studying Pedons

Pedons representative of an extensive mappable area are generally more useful than pedons that represent the border of an area or a small inclusion.

For a soil description to be of greatest value, the part of the landscape that the pedon represents and the vegetation should be described. This is referred to as the setting. The level of detail will depend on the objectives. A complete setting description should include information about the encompassing polypedon and, possibly, the polypedons conterminous with the encompassing polypedon. Furthermore, the setting may include information about the portion of the polypedon that differs from the central concept of the polypedon.

The description of a body of soil in the field, whether an entire pedon or a sample within it, should record the kinds of layers, their depth and thickness, and the properties of each layer. Generally, external features are observed throughout the extent of the polypedon; internal features are observed from the study of a pedon or that part of a pedon that is judged to be representative of the polypedon.

A pedon for detailed study of a soil is tentatively selected and then examined preliminarily to verify that it represents the desired segment of its range.

A pit exposing a vertical face approximately 1 meter across to an appropriate depth is satisfactory for most soils.

After the sides of the pit are cleaned of all loose material disturbed by digging, the exposed vertical faces are examined, usually starting at the top and working downward, to identify significant changes in properties. Boundaries between layers are marked on the face of the pit, and the layers are identified and described.

Photographs should be taken after the layers have been identified but before the vertical section is disturbed in the description-writing process. A point-count for estimation of the volume of stones or other features also is done before the layers are disturbed.

A horizontal view of each layer is useful. This exposes structural units that otherwise may not be observable. Patterns of colour within structural units, variations of particle size from the outside to the inside of structural units, and the pattern in which roots penetrate structural units are often seen more clearly in a horizontal section.

Excavations associated with roads, railways, gravel pits, and other soil disturbances provide easy access for studying soils; old exposures, however, must be used cautiously. The soils dry out or freeze and thaw from both the surface and the sides. Frequently, the soil structure in such excavations is more pronounced than is typical; salts may accumulate near the edges of exposures or be removed by seepage; and other changes may have taken place.

Depth to and Thickness of Horizons and Layers

Depth is measured from the soil surface. The soil surface is the top of the mineral soil; or, for soils with an 0 horizon, the soil surface is the top of the part of the 0 horizon that is at least slightly decomposed. Fresh leaf or needle fall that has not undergone observable decomposition is excluded from soil and may be described separately. The top of any surface horizon identified as an O horizon, whether Oi, Oe, or Oa, is considered the soil surface.

For soils with a cover of 80 percent or more rock fragments on the surface, the depth is measured from the surface of the rock fragments.

The depth to a horizon or layer boundary commonly differs within short distances, even within a pedon. The part of the pedon that is typical or most common is described. In the soil description, the horizon or layer designation is listed and is followed by the values that represent the depths from the soil surface to the upper and lower boundaries, in that order.

The depth to the lower boundary of a horizon or layer is the depth to the upper boundary of the horizon or layer beneath it. The variation in the depths of the boundaries is recorded in the description of the horizon or layer. The depth limits of the deepest horizon or layer described include only that part actually seen.

In some soils the variations in depths to boundaries are so complex that usual terms for description of topography of the boundary are inadequate. These variations are described separately. For example, "depth to the lower boundary is mainly 30 to 40 cm, but tongues extend to depths of 60 to 80 cm." The lower boundary of horizon or layer and the upper boundary of the horizon or layer below share a common irregularity.

The thickness of each horizon or layer is the vertical distance between the upper and lower boundaries. Thickness may vary within a pedon, and this variation should be shown in the description. A range in thickness may be given. It cannot be calculated from the range of upper and lower boundaries but rather must be evaluated across the exposure at different lateral points.

The location of upper and lower boundaries are commonly in different places. The upper boundary of a horizon, for example, may range in depth from 25 to 45 cm and the lower boundary from 50 to 75 cm. Taking the extremes of these two ranges, a wrong conclusion could be that the horizon ranges in thickness from as little as 5 cm to as much as 50 cm.

Land Surface Configuration

Land surface configuration considered here is geometrical and includes *soil slope* and *land surface shape*. Landform from a morphogenetic aspect is not considered. It may be applicable to a pedon or to a larger area.

Land surface configuration and relief are quite different as used here, although the meanings may be similar in other contexts. *Relief*, in this context, refers to the elevation or differences in elevation above mean sea level, considered collectively, of a land surface on a broad scale. Elevation can be determined from topographic maps or by using a calibrated altimeter.

Soil Slope

Slope has a scale connotation. It refers to the ground surface configuration for scales that exceed about 10 meters and range upward to the landscape as a whole. Slope has gradient, complexity, length, and aspect. The scale of reference commonly exceeds that of the pedon and should be indicated. The scale may embrace a map unit delineation, component of it, or an arbitrary area.

Slope gradient is the inclination of the surface of the soil from the horizontal. It is generally measured with a hand level. The difference in elevation between two points is expressed as a percentage of the distance between those points. If the difference in elevation is 1 meter over a horizontal distance of 100 meters, slope gradient is 1 percent. A slope of 45° is a slope of 100 percent, because the difference in elevation between two points 100 meters apart horizontally is 100 meters on a 45° slope.

Overland flow gradient is the slope of the soil surface in the direction of flow of surface water if it were present. The following examples show equivalences between percentage gradient and degree of slope angle:

Table 1:

Percentage	Angle	Angle	Percentage
0	0°00'	0°	0
5	2°52'	2°	3.5
10	5°43'	4°	7.0
15	8°32'	6°	10.5
20	11°19'	8°	14.0
25	14°02'	10°	17.6
30	16°42'	12°	21.2
35	19°17'	15°	26.8
40	21°48'	20°	36.4
50	26°34'	25°	46.6
60	30°58'	30°	57.7
70	34°59'	35°	70.0
80	38°39'	40°	83.9
90	41°59'	45°	100.0
100	45°00'	50°	119.2

Slope Complexity refers to surface form on the scale of a mapping unit delineation. In many places internal soil properties are more closely related to the slope complexity than to the gradient. Slope complexity has an important influence on the amount and rate of runoff and on sedimentation associated with runoff.

A guide to terminology for various slope classes defined in terms of gradient and complexity. The terms are used in discussing soil slope, and they can also be used in naming slope phases.

Table 2: Definitions of slope classes

Classes Simple Slopes	Complex Slopes	Slope gradient limits	
		Lower Percent	Upper Percent
Nearly level	Nearly level	0	3
Gently sloping	Undulating	1	8
Strongly sloping	Rolling	4	16
Moderately steep	Hilly	10	30
Steep	Steep	20	60
Very steep	Very steep	>45	

Terms are provided for both simple and for complex slopes in some classes. Complex slopes are groups of slopes that have definite breaks in several different directions and in most cases markedly different slope gradients within the areas delineated.

Significance of slope gradient is tied to other soil properties and to the purposes of soil surveys. Conventions are, therefore, provided in table 1 to adjust the slope limits of the various classes. Gently sloping or undulating soils, for example, can be defined to range as broadly as 1 to 8 percent or as narrowly as 3 to 5 percent. Classes may exceed the broadest range indicated in table 1 by a percentage point or two where the range is narrow and by as much as 5 percent or more where the range is broad.

If the detail of mapping requires slope classes that are more detailed than those in table 1, some of the classes can be divided as follows:

Nearly level: Level, Nearly level

Gently sloping: Very gently sloping, Gently sloping

Strongly sloping: Sloping, Strongly sloping, Moderately sloping

Undulating: Gently undulating, Undulating

Rolling: Rolling, Strongly rolling

In a highly detailed survey, for example, slope classes of 0 to 1 percent and 1 to 3 percent would be named "level" and "nearly level."

Slope length has considerable control over runoff and potential accelerated water erosion. Terms such as "long" or "short" can be used to describe slope lengths that are typical of certain kinds of soils. These terms are usually relative within a physiographic region. A "long" slope in one place might be "short" in another.

If such terms are used, they are defined locally. For observations at a particular point, it may be useful to record the length of the slope that contributes water to the point in addition to the total length of the slope. The former is called *point runoff slope length*. The *sediment transport slope length* is the distance from the expected or observed initiation upslope of runoff to the highest local elevation where deposition of sediment would be expected to occur. This distance need not be the same as the point runoff slope length.

Slope aspect is the direction toward which the surface of the soil faces. The direction is expressed as an angle between 0 degree and 360 degrees (measured clockwise from true north) or as a compass point such as east or north-northwest. Slope aspect may affect soil temperature, evapotranspiration, and winds received.

Land Surface Shape

Land surface shape has two components. One component is in a direction roughly parallel to the contours of the landform (or the contour lines on a map) as seen from directly overhead. The other component of shape is a direction perpendicular to the contours; that is, the shape of the slope as seen from the side. The shape parallel to the contours is less commonly consistent for a soil than is the shape perpendicular to the contours.

The shape parallel to the contours (across the slope) can be described by the shape of the contours. The shape is linear if contours are substantially a straight line, as on the side of a lateral moraine. An alluvial fan has a convex contour, as does a spur of the upland projecting into a valley. A cove on a hillside or a cirque in glaciated landscapes has concave contours. The two upper blocks have concave contours and the two lower blocks have convex contours. Where the contour is convex, runoff water tends to spread laterally as it moves down the slope. Where the contour is concave, runoff water tends to be concentrated toward the middle of the landform.

The shape of the surface at right angles to the contours (up and down the slope) may also be described as linear, convex, or concave.

Shape in this direction is usually identified simply as slope shape in contrast to slope contour in the other dimension. The surface of a linear slope is substantially a straight line when seen in profile at right angles to the contours. The gradient neither increases nor decreases significantly with distance. An example is the dip slope of a cuesta. On a concave slope, gradient decreases down the slope as on foot slopes. Runoff water tends to decelerate as it moves down the slope, and if it is loaded with sediment, the water tends to deposit the sediment on the lower parts of the slope. The soil on the lower part of the slope also tends to dispose of water less rapidly than the soil above it. On a convex slope, such as the shoulder or a ridge, gradient increases down the slope and runoff tends to accelerate as it flows down the slope. Soil on the lower part of the slope tends to dispose of water by runoff more rapidly than the soil above it. The soil on the lower part of a convex slope is subject to greater erosion than that on the higher part.

The configuration of the surface of a soil may be described in terms of both the shape of the contour and the shape of the slope. For example, a surface can be described as having a convex contour and a convex slope (an alluvial fan) or a linear contour and concave slope (the base of a moraine).

Microrelief refers to differences in ground-surface height, measured over distances of meters. Naturally formed features contrast with those that are tillage-determined. In areas of similar relief, the surface may be nearly uniform, or it may be interrupted by mounds, swales, or pits. Examples include the microrelief created when trees are blown over, referred to as cradle-knoll microrelief. This consists of the knoll left by the earth that clung to the roots of the tree when it was uprooted and the depression from which it came. Coppice dunes form where windblown soil material accumulates around widely spaced plants in arid regions. Gilgai produced by expansion and contraction of soils is a form of microrelief. Mima mounds and biscuit-scabland are other examples of microrelief, although individual mounds may cover 100 square meters or more.

Descriptions should indicate whether mounds or depressions are closed, form a network, or are in a linear pattern. If mounds rest on a smooth surface, their size and spacing should be described. At a specific site within an area having microrelief, it is important to note whether a described pedon is at a high point, on a slope, in a depression, or at some combination of these places. Internal soil properties in mounds may be different from the properties in depressions.

Roughness refers to a ground surface configuration with a repeat distance between prominences of less than 50 cm and for areas less than about 10 m across. This scale applies to most tillage operations and affects aspects of land surface water flow such as detention, infiltration, runoff, and erosion. Roughness, as used here, pertains to the ground surface and includes rock fragments on the surface. It does not include vegetation. If vegetation is included, the fact should be indicated. Roughness along a line, referred to as *one-dimensional roughness*, can be measured more easily than can roughness for an area.

Area measurements, however, permit the separation of random and tillage-determined roughness. The orientation to which the observation of *one-dimensional roughness* pertains must be specified relative to the direction of surface runoff or of air movement. Position within the tillage-determined relief, if present, should be indicated for *one-dimensional roughness*. An example of such a position would be the non-traffic interrow in a tilled field. The *standard deviation* of the ground surface height is the primary descriptor. There are a number of approaches to the measurement of roughness, and those who are in agronomic disciplines should be consulted. The measurements depend on the variation in height from a levelled reference. Photographs may be used to illustrate the classes; placement in classes may be made directly from the photographs.

Vegetation

Correlations between vegetation and soils are made for three main purposes: (1) understanding soil genesis, (2) recognizing soil boundaries, and (3) making predictions from soil maps about the kind and amount of vegetation produced.

The principal kinds of plants present are listed in order of their abundance. In annual cropland, the plant or plants that have been grown should be recorded, including significant weeds. In forested areas, separate treatment is often necessary for forest trees, understory of small trees and shrubs, and the ground cover. Many soils in range have an overstory of shrubs or low trees. These are listed separately from the grasses, forbs, and other ground cover. An idea of the density of stand or plant cover, such as average canopy cover of trees or shrubs, should be given. The range in size of dominant species of trees can be given as "diameter breast height," if desired. Estimated percentage of the ground covered by grasses and forbs should be included.

Common names of the plants may be used, if such names are clear and specific. In areas where the plants are important for the use and interpretation of the soil map, the soil survey record should include both common and scientific names of plants.

If possible, the kinds and amounts of plants in the potential natural vegetation on a soil should be estimated. This vegetation is closely related to the soil and its genesis. Generally, a close relationship exists between native vegetation and kinds of soil, yet there are important exceptions. Observations of the growth of native vegetation and cultivated crops aid in recognizing soil boundaries and provide direct information about the behaviour of specific plants on different kinds of soil. Within fields of a single crop, differences of vigour, stand, or colour of the crop or of weeds commonly mark soil differences and are valuable clues to the location of soil boundaries.

By studying many sites of the same kind of soil under different land-use history, the potential plant community and principles of plant succession for that kind of soil can be ascertained, particularly if range and forestry specialists provide assistance. Farmers learn which crops do well and which do poorly on different kinds of soil and adjust their cropping patterns accordingly. If the differences are large— as between crop failure and reasonable performance—the near absence of a given crop on a specific kind of soil questions the suitability of that kind of soil for the crop. If the differences are small, many non-soil factors can determine the farmer's choice of fields for a given crop. Yield information for cultivated crops, range, and trees should be associated with pedon descriptions insofar as possible.

Ground Surface Cover

The ground surface of most soils is covered to some extent at least part of the year by vegetation. Furthermore, in many soils rock fragments form part of the mineral material at the soil surface. Together, the vegetal material that is not part of the surface horizon and the rock fragments form the ground surface cover. The proportion of cover, together with its characteristics, is very important in determining thermal properties and resistance to erosion.

At one extreme, estimation of cover can be made visually without quantitative measurement. At the other extreme, transect techniques can be used to make a rather complete modal analyses of the ground surface. More effort is justified on ground surface documentation if it is relatively permanent. In many instances, a combination of rapid visual estimates and transect techniques is appropriate.

The ground surface may be divided into fine earth and material other than fine earth. The latter consists of rock fragments and both alive and dead vegetation. Vegetation is separated into *canopy* and *noncanopy*. A canopy component has a relatively large cross sectional area capable of intercepting rainfall compared to the area near enough to the ground surface to affect overland water flow. In practice, the separation of canopy from noncanopy should be coordinated with the protocols for computation of susceptibility to erosion. Noncanopy material is commonly referred to as *mulch*. It includes rock fragments and vegetation.

The first step in evaluation is to decide upon the ground surface cover components. The number is usually one to three. A common three-component land surface consists of trees, bushes, and areas between the two. The areal proportion of each component must be established. This may be done by transect. If a canopy component is present, the area within the drip line as a percent of the ground surface is determined.

For each canopy component, the effectiveness must be established. *Effectiveness* is the percent of vertical raindrops that would be intercepted. Usually the canopy effectiveness is estimated visually, but a spherical densitometer may be used. In addition to the canopy effectiveness, the mulch (rock fragments plus vegetation) must be established for each component.

Transect techniques may be employed to determine the mulch percentage. The mulch can be subdivided into rock fragments and vegetation. From the areal proportions of the components and their respective canopy efficiencies and mulch percentages, the soil-loss ratio may be computed for the whole land surface (Wischmeier, 1978). In addition to the observations for the computation of the soil-loss ratio, information may be obtained about the percent of kinds of plants, size of rock fragments, amount of green leaf area, and aspects of colour of the immediate surface that would affect absorption of radiant energy in an area.

Parent Material

Parent material refers to unconsolidated organic and mineral materials in which soils form. The parent material of a genetic horizon cannot be observed in its original state; it must be inferred from the properties that the horizon has inherited and from other evidence. In some soils, the parent material has changed little, and what it was like can be deduced with confidence. In others, such as some very old soils of the tropics, the specific kind of parent material or its mode or origin is speculative.

Much of the mineral matter in which soils form is derived in one way or another from hard rocks. Glaciers may grind the rock into fragments and earthy material and deposit the mixture of particles as glacial till. On the other hand, rock may be weathered with great chemical and physical changes but not moved from its place of origin; this altered material is called "residuum from rock."

In some cases, little is gained from attempting to differentiate between geologic weathering and soil formation because both are weathering processes. It may be possible to infer that a material was weathered before soil formation. The weathering process causes some process constituents to be lost, some to be transformed, and others to be concentrated.

Parent material may not necessarily be residuum from the bedrock that is directly below, and the material that developed into a modern soil may be unrelated to the underlying bedrock. Movement of soil material downslope is an important process and can be appreciable even on gentle slopes, especially on very old landscapes. Also, locally associated soils may form in sedimentary rock layers that are different.

Seldom is there certainty that a highly weathered material weathered in place. The term "residuum" is used when the properties of the soil indicate that it has been derived from rock like that which underlies it and when evidence is lacking that it has been modified by movement. A rock fragment distribution that decreases in amount with depth, especially over saprolite, indicates that soil material probably has been transported downslope.

Stone lines, especially if the stones have a different lithology than the underlying bedrock, provide evidence that the soil did not form entirely in residuum. In some soils, transported material overlies residuum and illuvial organic matter and clay are superimposed across the discontinuity between the contrasting materials. A certain degree of landscape stability is inferred for residual soils. A lesser degree is inferred for soils that developed in transported material.

Both consolidated and unconsolidated material beneath the solum that influence the genesis and behaviour of the soil are described in standard terms. Besides the observations themselves, the scientist records his judgment about the origin of the parent material from which the solum developed. The observations must be separated clearly from inferences.

The lithologic composition, structure and consistence of the material directly beneath the solum are important. Evidence of stratification of the material—textural differences, stone lines, and the like—need to be noted. Commonly, the upper layers of outwash deposits settled out of more slowly moving water and are finer in texture than the lower layers. Windblown material and volcanic ash are laid down at different rates in blankets of varying thickness. Examples of such complications are nearly endless.

Where alluvium, loess, or ash are rapidly deposited on old soils, buried soils may be well preserved. Elsewhere the accumulation is so slow that the solum thickens only gradually. In such places, the material beneath the solum was once near the surface but may now be buried below the zone of active change.

Where hard rocks or other strongly contrasting materials lie near enough to the surface to affect the behaviour of the soil, their depths need to be measured accurately. The depth of soil over such nonconforming materials is an important criterion for distinguishing different kinds of soil.

Geological materials need to be defined in accordance with the accepted standards and nomenclature of geology. The accepted, authoritative names of the geological formations are recorded in soil descriptions where these can be identified with reasonable accuracy. As soil research progresses, an increasing number of correlations are being found between particular geological formations and the mineral and nutrient content of parent materials and soils. For example, certain terrace materials and deposits of volcanic ash that are different in age or source, but otherwise indistinguishable, vary widely in the content of cobalt. Wide variations in the phosphorus content of two otherwise similar soils may reflect differences in the phosphorus content of two similar limestones that can be distinguished in the field only by specific fossils.

Igneous rocks formed by the solidification of molten materials that originated within the earth. Examples of igneous rocks that weather to important soil material are granite, syenite, basalt, andesite, diabase, and rhyolite.

Sedimentary rocks formed from sediments laid down in previous geological ages. The principal broad groups of sedimentary rocks are limestone, sandstone, shale, and conglomerate. There are many varieties of these broad classes of sedimentary rocks; for example, chalk and marl are soft varieties of limestone. Many types are intermediate between the broad groups, such as calcareous sandstone and arenaceous limestone. Also included are deposits of diatomaceous earth, which formed, from the siliceous remains of primitive plants called diatoms.

Metamorphic rocks resulted from profound alteration of igneous and sedimentary rocks by heat and pressure. General classes of metamorphic rocks important as parent material are gneiss, schist, slate, marble, quartzite, and phyllite.

The principal broad subdivisions of parent material are discussed in the following paragraphs.

Material Produced by Weathering of Rock in Place

The nature of the original rock affects the kinds of material produced by weathering. The rock may have undergone various changes, including changes in volume and loss of minerals—plagioclase feldspar and other minerals. Rock may lose mineral material without any change in volume or in the original rock structure, and saprolite is formed. Essentially, saprolite is a parent material. The point where rock weathering ends and soil formation begins is not always clear. The processes may be consecutive and even overlapping. Quite different soils may form from similar or even identical rocks under different weathering conditions. Texture, colour, consistence, and other characteristics of the material should be included in the description of soils, as well as important features such as quartz dikes. Useful information about the mineralogical composition, consistence, and structure of the parent rock itself should be added to help in understanding the changes from parent rock to weathered material.

Transported Material

The most extensive group of parent materials is the group that has been moved from the place of origin and deposited elsewhere. The principal groups of transported materials are usually named according to the main agent responsible for their transport and deposition. In most places, sufficient evidence is available to make a clear determination; elsewhere, the precise origin is uncertain.

In soil morphology and classification, it is exceedingly important that the characteristics of the material itself be observed and described. It is not enough simply to identify the parent material. Any doubt of the correctness of the identification should be mentioned.

For example, it is often impossible to be sure whether certain silty deposits are alluvium, loess, or residuum. Certain mud flows are indistinguishable from glacial till. Some sandy glacial till is nearly identical to sandy outwash. Fortunately, hard-to-make distinctions are not always of significance for soil behaviour predictions.

Material Moved and Deposited by Water

Alluvium: Alluvium consists of sediment deposited by running water. It may occur on terraces well above present streams or in the normally flooded bottom land of existing streams. Remnants of very old stream terraces may be found in dissected country far from any present stream. Along many old established streams lie a whole series of alluvial deposits in terraces—young deposits in the immediate flood plain, up step by step to the very old deposits on the highest terraces. In some places recent alluvium covers older terraces.

Lacustrine deposits: These deposits consist of material that has settled out of bodies of still water. Deposits laid down in freshwater lakes associated directly with glaciers are commonly included as are other lake deposits, including some of Pleistocene age that are not associated with the continental glaciers. Some lake basins in the Western United States are commonly called playas; the soils in these basins may be more or less salty, depending on climate and drainage.

Marine sediments: These sediments settled out of the sea and commonly were reworked by currents and tides. Later they were exposed either naturally or following the construction of dikes and drainage canals. They vary widely in composition. Some resemble lacustrine deposits.

Beach deposits: Beach deposits mark the present or former shorelines of the sea or lakes. These deposits are low ridges of sorted material and are commonly sandy, gravelly, cobbly, or stony. Deposits on the beaches of former glacial lakes are usually included with glacial drift.

Material Moved and Deposited by Wind

Windblown material can be divided into groups based on particle size or on origin. Volcanic ash and cinders are examples of materials classed by both particle size and origin. Other windblown material that is mainly silty is called loess, and that which is primarily sand is called eolian sand. Eolian sand is commonly but not always in dunes. Nearly all textures intermediate between silty loess and sandy dune material can be found.

Volcanic ash, pumice, and *cinders* are sometimes regarded as unconsolidated igneous rock, but they have been moved from their place of origin. Most have been reworked by wind and, in places, by water. Ash is volcanic ejecta smaller than 2 mm. Ash smaller than 0.05 mm may be called "fine ash." Pumice and cinders are volcanic ejecta 2 mm or larger.

Loess deposits typically are very silty but may contain significant amounts of clay and very fine sand. Most loess deposits are pale brown to brown, although gray and red colours are common. The thick deposits are generally massive and have some gross vertical cracking. The walls of road cuts in thick loess stand nearly vertical for years. Other silty deposits that formed in other ways have some or all of these characteristics. Some windblown silt has been leached and strongly weathered so that it is acid and rich in clay. On the other hand, some young deposits of windblown material (loess) are mainly silt and very fine sand and are low in clay.

Sand dunes, particularly in warm, humid regions, characteristically consist of fine or medium sand that is high in quartz and low in clay-forming materials. Sand dunes may contain large amounts of calcium carbonate or gypsum, especially in deserts and semideserts.

During periods of drought and in deserts, local wind movements may mix and pile up soil material of different textures or even material that is very rich in clay. Piles of such material have been called "soil dunes" or "clay dunes." Rather than identify local accumulations of mixed material moved by the wind as "loess" or "dunes," however, it is better to refer to them as "wind-deposited material."

Also important but not generally recognized as a distinctive deposit is *dust*, which is carried for long distances and deposited in small increments on a large part of the world. Dust can circle the earth in the upper atmosphere. Dust particles are mostly clay and very fine silt and may be deposited dry or be in precipitation. The accumulated deposits are large in some places. An immense amount of dust has been distributed widely throughout the ages. The most likely sources at present are the drier regions of the world. Large amounts of dust may have been distributed worldwide during and immediately following the glacial periods.

Dust is an important factor affecting soils in some places. It is the apparent source of the unexpected fertility of some old, highly leached soils in the path of wind that blows from extensive deserts some hundreds of kilometers distant. It explains unexpected micronutrient distribution in some places. Besides dust, fixed nitrogen, sulfur, calcium, magnesium, sodium, potassium, and other elements from the atmosphere are deposited on the soil in varying amounts in solution in precipitation.

Material Moved and Deposited by Glacial Processes

Several terms are used for material that has been moved and deposited by glacial processes. *Glacial drift* consists of all of the material picked up, mixed, disintegrated, transported, and deposited by glacial ice or by water from melting glaciers. In many places glacial drift is covered by a mantle of loess. Deep mantles of loess are usually easily recognized, but very thin mantles may be so altered by soil-building forces that they can scarcely be differentiated from the underlying modified drift.

Glacial till: This is that part of the glacial drift deposited directly by the ice with little or no transportation by water. It is generally an unstratified, heterogeneous mixture of clay, silt, sand, gravel, and sometimes boulders. Some of the mixture settled out as the ice melted with very little washing by water, and some was overridden by the glacier and is compacted and unsorted. Till may be found in ground moraines, terminal moraines, medial moraines, and lateral moraines. In many places it is important to differentiate between the tills of the several glaciations. Commonly, the tills underlie one another and may be separated by other deposits or old, weathered surfaces. Many deposits of glacial till were later eroded by the wave action in glacial lakes. The upper part of such wave-cut till may have a high percentage of rock fragments.

Glacial till ranges widely in texture, chemical composition, and the degree of weathering that followed its deposition. Much till is calcareous, but an important part is noncalcareous because no carbonate rocks contributed to the material or because subsequent leaching and chemical weathering have removed the carbonates.

Glaciofluvial deposits: These deposits are material produced by glaciers and carried, sorted, and deposited by water that originated mainly from melting glacial ice. *Glacial outwash* is a broad term for material swept out, sorted, and deposited beyond the glacial ice front by streams of melt water. Commonly, this outwash is in the form of plains, valley trains, or deltas in old glacial lakes. The valley trains of outwash may extend far beyond the farthest advance of the ice. Near moraines, poorly sorted glaciofluvial material may form kames, eskers, and crevasse fills.

Glacial beach deposits: These consist of rock fragments and sand. They mark the beach lines of former glacial lakes. Depending on the character of the original drift, beach deposits may be sandy, gravelly, cobbly, or stony.

Glaciolacustrine deposits: These deposits are derived from glaciers but were reworked and laid down in glacial lakes. They range from fine clay to sand. Many of them are stratified or varved. A *varve* consists of the deposition for a calendar year. The finer portion reflects slower deposition during the cold season and the coarser portion deposition during the warmer season when runoff is greater.

Good examples of all of the glacial materials and forms described in the preceding paragraphs can be found. In many places, however, it is not easy to distinguish definitely among the kinds of drift on the basis of mode of origin and landform. For example, pitted outwash plains can scarcely be distinguished from sandy till in terminal moraines. Distinguishing between wave-cut till and lacustrine material is often difficult. The names themselves connote only a little about the actual characteristics of the parent material.

Material Moved and Deposited by Gravity

Colluvium is poorly sorted debris that has accumulated at the base of slopes, in depressions, or along small streams through gravity, soil creep, and local wash. It consists largely of material that has rolled, slid or fallen down the slope under the influence of gravity. Accumulations of rock fragments are called talus. The rock fragments in colluvium are usually angular, in contrast to the rounded, water-worn cobbles and stones in alluvium and glacial outwash.

Organic Material

Organic material accumulates in wet places where it is deposited more rapidly than it decomposes. These deposits are called peat. This peat in turn may become parent material for soils. The principal general kinds of peat, according to origin are:

Sedimentary peat. the remains mostly of floating aquatic plants, such as algae, and the remains and fecal material of aquatic animals, including coprogenous earth.

Moss peat. the remains of mosses, including Sphagnum.

Herbaceous peat. the remains of sedges, reeds, cattails, and other herbaceous plants.

Woody peat. the remains of trees, shrubs, and other woody plants.

Many deposits of organic material are mixtures of peat. Some organic soils formed in alternating layers of different kinds of peat. In places peat is mixed with deposits of mineral alluvium and/or volcanic ash. Some organic soils contain layers that are largely or entirely mineral material.

In describing organic soils, the material is called *peat* (fibric) if virtually all of the organic remains are sufficiently fresh and intact to permit identification of plant forms. It is called *muck* (sapric) if virtually all of the material has undergone sufficient decomposition to limit recognition of the plant parts. It is called *mucky peat* (hemic) if a significant part of the material can be recognized and a significant part cannot.

Descriptions of organic material should include the origin and the botanical composition of the material to the extent that these can be reasonably inferred.

Contrasting Materials

Changes with depth that are not primarily related to pedogenesis but rather to geological processes are *contrasting soil materials* if they are sufficient to affect use and management. The term *discontinuity* is applied to certain kinds of contrasting soil materials.

Unconsolidated contrasting soil material may differ in pore-size distribution, particle-size distribution, mineralogy, bulk density, or other properties. Some of the differences may not be readily observable in the field. Some deposits are clearly stratified, such as some lake sediments and glacial outwash, and the discontinuities may be sharply defined.

Contrasting materials can be confused with the effects of soil formation. Silt content may decrease regularly with depth in soils presumed to have formed in glacial till. The higher silt content in the upper part of these soils can be explained by factors other than soil formation. In some of these soils, small amounts of eolian material may have been deposited on the surface over the centuries and mixed by insects and rodents with the underlying glacial till. In others, the silt distribution reflects water sorting.

Inferences about contrasting properties inherited from differing layers of geologic material may be noted when the soil is described. Generally, each identifiable layer that differs clearly in properties from adjacent layers is recognized as a subhorizon. Whether it is recognized as a discontinuity or not depends on the degree of contrast with overlying and underlying layers and the thickness. For many soils the properties inherited from even sharply contrasting layers are not consistent from place to place and are described in general terms. The C layer of a soil in stratified lake sediments, for example, might be described as follows: "consists of layers of silt and clay, 1 to 20 cm thick; the aggregate thickness of layers of silt and that of the layers of clay are in a ratio of about 4 to 1; material is about 80 percent silt."

Erosion

Erosion is the detachment and movement of soil material. The process may be natural or accelerated by human activity. Depending on the local landscape and weather conditions, erosion may be very slow or very rapid.

Natural erosion has sculptured landforms on the uplands and built landforms on the lowlands. Its rate and distribution in time controls the age of land surfaces and many of the internal properties of soils on the surfaces. The formation of Channel Scablands in the state of Washington is an example of extremely rapid natural, or geologic, erosion. The broad, nearly level interstream divides on the Coastal Plain of the Southeastern United States are examples of areas with very slow or no natural erosion.

Landscapes and their soils are evaluated from the perspective of their natural erosional history. Buried soils, stone lines, deposits of wind-blown material, and other evidence that material has been moved and redeposited is helpful in understanding natural erosion history. Thick weathered zones that developed under earlier climatic conditions may have been exposed to become the material in which new soils formed. In landscapes of the most recently glaciated areas, the consequences of natural erosion, or lack of it, are less obvious than where the surface and the landscape are of an early Pleistocene or even Tertiary age. Even on the landscapes of most recent glaciation, however, postglacial natural erosion may have redistributed soil materials on the local landscape. Natural erosion is an important process that affects soil formation and, like man-induced erosion, may remove all or part of soils formed in the natural landscape.

Accelerated erosion is largely the consequence of human activity. The primary causes are tillage, grazing, and cutting of timber.

The rate of erosion can be increased by activities other than those of humans. Fire that destroys vegetation and triggers erosion has the same effect. The spectacular episodes of erosion, such as the soil blowing on the Great Plains of the Central United States in the 1930s, have not all been due to human habitation. Frequent dust storms were recorded on the Great Plains before the region became a grain-producing area. "Natural" erosion is not easily distinguished from "accelerated" erosion on every soil. A distinction can be made by studying and understanding the sequence of sediments and surfaces on the local landscape, as well as by studying soil properties.

Landslip Erosion

Landslip erosion refers to the mass movement of soil. Slides and flows are two kinds of landslip erosion. In the slide process, shear takes place along one or a limited number of surfaces. Slide movement may be categorized as *slightly* or *highly deformed*, depending on the extent of rearrangement from the original organization. In flow movement the soil mass acts as a viscous fluid. Failure is not restricted to a surface or a small set of surfaces. Classes of landslip erosion are not provided. Location of the mass movement relevant to landscape features generally and the size of the mass movement in terms of area parallel to the land surface and the depth may be indicated. Information about the time since the mass movement took place may be very useful.

Water Erosion

Water erosion results from the removal of soil material by flowing water. A part of the process is the detachment of soil material by the impact of raindrops. The soil material is suspended in runoff water and carried away. Four kinds of accelerated water erosion are commonly recognized: sheet, rill, gully, and tunnel (piping).

Sheet erosion is the more or less uniform removal of soil from an area without the development of conspicuous water channels. The channels are tiny or tortuous, exceedingly numerous, and unstable; they enlarge and straighten as the volume of runoff increases. Sheet erosion is less apparent, particularly in its early stages, than other types of erosion. It can be serious on soils that have a slope gradient of only 1 or 2 percent; however, it is generally more serious as slope gradient increases.

Rill erosion is the removal of soil through the cutting of many small, but conspicuous, channels where runoff concentrates. Rill erosion is intermediate between sheet and gully erosion. The channels are shallow enough that they are easily obliterated by tillage; thus, after an eroded field has been cultivated, determining whether the soil losses resulted from sheet or rill erosion is generally impossible.

Gully erosion is the consequence of water that cuts down into the soil along the line of flow. Gullies form in exposed natural drainageways, in plow furrows, in animal trails, in vehicle ruts, between rows of crop plants, and below broken man-made terraces. In contrast to rills, they cannot be obliterated by ordinary tillage. Deep gullies cannot be crossed with common types of farm equipment.

Gullies and gully patterns vary widely. V-shaped gullies form in material that is equally or increasingly resistant to erosion with depth. U-shaped gullies form in material that is equally or decreasingly resistant to erosion with depth. As the substratum is washed away, the overlying material loses its support and falls into the gully to be washed away. Most-U-shaped gullies become modified toward a V shape once the channel stabilizes and the banks start to spall and slump.

The maximum depth to which gullies are cut is governed by resistant layers in the soil, by bedrock, or by the local base level. Many gullies develop headward; that is, they extend up the slope as the gully deepens in the lower part.

Tunnel erosion may occur in soils with subsurface horizons or layers that are more subject to entrainment in moving free water than is the surface horizon or layer. The free water enters the soil through ponded infiltration into surface-connected macropores. Desiccation cracks and rodent burrows are examples of macropores that may initiate the process. The soil material entrained in the moving water moves downward within the soil and may move out of the soil completely if there is an outlet.

The result is the formation of tunnels (also referred to as pipes) which enlarge and coalesce. The portion of the tunnel near the inlet may enlarge disproportionately to form a funnel-shaped feature often referred to as a "jug." Hence, the term "piping" and "jugging." The phenomenon is favoured by the presence of appreciable exchangeable sodium.

Deposition of sediment carried by water is likely anywhere that the velocity of running water is reduced—at the mouth of gullies, at the base of slopes, along stream banks, on alluvial plains, in reservoirs, and at the mouth of streams. Rapidly moving water, when slowed, drops stones, then cobbles, pebbles, sand, and finally silt and clay. Sediment transport slope length has been defined as the distance from the highest point on the slope where runoff may start to where the sediment in the runoff would be deposited.

Wind Erosion

Wind Erosion in regions of low rainfall, can be widespread, especially during periods of drought. Unlike water erosion, wind erosion is generally not related to slope gradient. The hazard of wind erosion is increased by removing or reducing the vegetation.

When winds are strong, coarser particles are rolled or swept along on or near the soil surface, kicking finer particles into the air. The particles are deposited in places sheltered from the wind. When wind erosion is severe, the sand particles may drift back and forth locally with changes in wind direction while the silt and clay are carried away. Small areas from which the surface layer has blown away may be associated with areas of deposition in such an intricate pattern that the two cannot be identified separately on soil maps.

Estimating the Degree of Erosion

The degree to which accelerated erosion has modified the soil may be estimated during soil examinations. The conditions of eroded soil are based on a comparison of the suitability for use and the management needs of the eroded soil with those of the uneroded soil. The eroded soil is identified and classified on the basis of the properties of the soil that remains. An estimate of the soil lost is described. Eroded soils are defined so that the boundaries on the soil maps separate soil areas of unlike use suitabilities and unlike management needs.

The depth to a reference horizon or soil characteristic of the soil under a use that has minimized accelerated erosion are compared to the same properties under uses that have favoured accelerated erosion. For example, a soil that supports native grass or large trees with no evidence of cultivation would be the basis for comparison of the same or similar soil that has been cleared and cultivated for a relatively long time.

The depth to reference layers is measured from the top of the mineral soil because organic horizons at the surface of mineral soils are destroyed by cultivation.

The depths to a reference layer must be interpreted in terms of recent soil use or history. Cultivation may cause differences in thickness of layers. The upper parts of many forested soils have roots that make up as much as one-half of the soil volume. When these roots decay, the soil settles. Rock fragment removal can also lower the surface. The thickness of surficial zones that have been bulked by tillage should be adjusted downward to what they would be if water had compacted them.

The thickness of a plowed layer of a specific soil cannot be used as a standard for either losses or additions of material because, as a soil erodes, the plow cuts progressively deeper. Nor can the thickness of the uncultivated and uneroded A horizon be used as a standard for all cultivated soil, unless the A horizon is much thicker than the plow layer.

If the horizon immediately below the plowed layer of an uneroded soil is distinctly higher in clay than the A horizon, the plow layer becomes progressively more clayey under continued cultivation as erosion progresses; the texture of the plow layer may then be a criterion of erosion.

Comparisons must be made on comparable slopes. Near the upper limit of the range of slope gradient for a soil, horizons may normally be thinner than near the lower limit of the range for the same soil.

Roadsides, cemeteries, fence rows, and similar uncultivated areas that are a small part of the landscape as a whole or are subject to unusual cultural histories must be used cautiously for setting standards, because the reference standards for surface-layer thickness are generally set too high. In naturally treeless areas or in areas cleared of trees, dust may collect in fence rows, along roadsides, and in other small uncultivated areas that are covered with grass or other stabilizing plants. The dust thus accumulated may cause the surface horizon to become several centimetres thicker in a short time.

For soils having clearly defined horizons, differences due to erosion can be accurately determined by comparison of the undisturbed or uncultivated norms within the limitations discussed. Guides for soils having a thin A horizon and little or no other horizon are more difficult to establish. After the thin surface layer is gone or has been mixed with underlying material, few clues remain for estimating the degree of erosion. The physical conditions of the material in the plowed layer, the appearance and amount of rock fragments on the surface, the number and shape of gullies, and similar evidence are relied on. For many soils having almost no horizon expression, attempting to estimate the degree of erosion serves little useful purpose.

Classes of Accelerated Erosion

The classes of accelerated erosion that follow apply to both water and wind erosion. They are not applicable to landslip or tunnel erosion. The classes pertain to the proportion of upper horizons that have been removed. These horizons may range widely in thickness; therefore, the absolute amount of erosion is not specified.

Class 1. This class consists of soils that have lost some, but on the average less than 25 percent, of the original A and/or E horizons or of the uppermost 20 cm if the original A and/or E horizons were less than 20 cm thick. Throughout most of the area, the thickness of the surface layer is within the normal range of variability of the uneroded soil. Scattered small areas amounting to less than 20 percent of the area may be modified appreciably. Evidence for class 1 erosion includes (1) a few rills, (2) an accumulation of sediment at the base of slopes or in depressions, (3) scattered small areas where the plow layer contains material from below, and (4) evidence of the formation of widely spaced, deep rills or shallow gullies without consistently measurable reduction in thickness or other change in properties between the rills or gullies.

Class 2. This class consists of soils that have lost, on the average, 25 to 75 percent of the original A and/or E horizons or of the uppermost 20 cm if the original A and/or E horizons were less than 20 cm thick. Throughout most cultivated areas of class 2 erosion, the surface layer consists of a mixture of the original A and/or E horizons and material from below. Some areas may have intricate patterns, ranging from uneroded small areas to severely eroded small areas. Where the original A and/or E horizons were very thick, little or no mixing of underlying material may have taken place.

Class 3. This class consists of soils that have lost, on the average, 75 percent or more of the original A and/or E horizons or of the uppermost 20 cm if the original A and/or E horizons were less than 20 cm thick. In most areas of class 3 erosion, material below the original A and/or E horizons is exposed at the surface in cultivated areas; the plow layer consists entirely or largely of this material. Even where the original A and/or E horizons were very thick, at least some mixing with underlying material generally took place.

Class 4. This class consists of soils that have lost all of the original A and/or E horizons or the uppermost 20 cm if the original A and/or E horizons were less than 20 cm thick. In addition, Class 4 includes some or all of the deeper horizons throughout most of the area. The original soil can be identified only in small areas. Some areas may be smooth, but most have an intricate pattern of gullies.

Soil Water

This section discusses "the water regime"—schemes for the description of the state of the soil water at a particular time and for the change in soil water state over time. Soil water state is evaluated from water suction, quantity of water, whether the soil water is liquid or frozen, and the occurrence of free water within the soil and on the land surface. Complexity and detail of water regime statements may range widely.

Inundation Classes

Free water may occur above the soil. Inundation is the condition that the soil area is covered by liquid free water. Flooding is temporary inundation by flowing water. If the water is standing, as in a closed depression, the term ponding is used.

Internal Classes

Definitions: Table 2 contains water state classes for the description of individual layers or horizons. Only matrix suction is considered in definition of the classes. Osmotic potential is not considered. For water contents of medium and fine-textured soil materials at suctions less than about 200 kPa, the reference laboratory water retention is for the natural soil fabric. Class limits are expressed both in terms of suction and water content. In order to make field and field office evaluation more practicable, water content pertains to gravimetric quantities and not to volumetric. The classes are applicable to organic as well as to mineral soil material. The frozen condition is indicated separately by the symbol "f." The symbol indicates the presence of ice; some of the water may not be frozen. If the soil is frozen, the water content or suction pertains to what it would be if not frozen.

Table 3. Water state classes

Class	Criteria[a]
Dry (D)	>1500 kPa suction
Very Dry (DV)	<(0.35 x 1500 kPa retention)
Moderately Dry (DM)	0.35 to 0.8 x 1500 kPa retention
Slightly Dry (DS)	0.8 to 1.0 x 1500 kPa retention
Moist (M)	<1500 kPa to >1 or 1/2 kPa[b]
Slightly Moist (MS)	1500 kPa suction to MWR[c]
Moderately Moist (MM)	MWR to UWR[c]
Very Moist (MV)	UWR to 1 or 1/2 kPa[b] suction
Wet <1 kPa or <1/2 kPa[b]	
Nonsatiated (WN)	No free water
Satiated (WA)	Free water present

a. Criteria use both suction and gravimetric water contents as defined by suction.b. 1/2 kPa only if coarse soil material.c. UWR is the abbreviation for upper water retention, which is the laboratory water retention at 5 kPa for coarse soil material and 10 kPa for other. MWR is the midpoint water retention. It is halfway between the upper water retention and the retention at 1500 kPa.

Three classes and eight subclasses are defined. Classes and subclasses may be combined as desired. Symbols for the combinations currently defined are in table 2. Specificity desired and characteristics of the water desorption curve would determine whether classes or subclasses would be used. Coarse soil material has little water below the 1500 kPa retention, and so subdivisions of dry generally would be less useful.

Dry is separated from *moist* at 1500 kPa suction. *Wet* is separated from moist at the condition where water films are readily apparent. The water suction at the moist-wet boundary is assumed to be about 1/2 kPa for coarse soil materials and 1 kPa for other materials. The formal definition of coarse soil material is given later.

Three subclasses of dry are defined—*very dry, moderately dry*, and *slightly dry*. Very dry cannot be readily distinguished from air dry in the field. The water content extends from ovendry to 0.35 times the water retention at 1500 kPa. The upper limit is roughly 150 percent of the air dry water content. The limit between moderately dry and slightly dry is a water content 0.8 times the retention at 1500 kPa.

The moist class is subdivided into *slightly moist, moderately moist*, and *very moist*. Depending on the kind of soil material, laboratory retention at 5 or 10 kPa suction (method 4B, Soil Survey Laboratory Staff, 1992) determines the *upper water retention*. A suction of 5 kPa is employed for coarse soil material. Otherwise, 10 kPa is used.

To be considered coarse, the soil material that is strongly influenced by volcanic ejecta must be nonmedial and weakly or nonvesicular. If not strongly influenced by volcanic ejecta, it must meet the sandy or sandy-skeletal family particle size criteria and also be coarser than loamy fine sand, have <2 percent organic carbon, and have <5 percent water at 1500 kPa suction. Furthermore, the computed total porosity of the <2 mm fabric must exceed 35 percent.

Very moist has an upper limit at the moist-wet boundary and a lower limit at the upper water retention. Relatedly, *moderately moist* has an upper limit at the upper water retention and a lower limit at the midpoint in gravimetric water content between retention at 1500 kPa and the upper water retention. This lower limit is referred to as the midpoint water retention. *Slightly moist*, in turn, extends from the midpoint water retention to the 1500 kPa retention.

The wet class has *nonsatiated* and *satiated* subclasses distinguished on the basis of absence or presence of free water. Miller and Bresler (1977) defined satiation as the condition from the first appearance of free water through saturation. The nonsatiated wet state may be applicable at zero suction to horizons with low or very low saturated hydraulic conductivity. These horizons may not exhibit free water. Horizons may have parts that are *satiated wet* and other parts, because of low matrix saturated hydraulic conductivity and the absence of conducting macroscopic pores, that are *nonsatiated wet*. Free water develops positive pressure with depth below the top of a wet satiated zone.

A class for saturation (that is, zero air-filled porosity) is not provided because the term suggests that all of the pore space is filled with water. This condition usually cannot be evaluated in the field. Further, if saturation is used for the concept of satiation, then a term is not available to describe known saturation. There is an implication of saturation if the soil material is satiated wet and coarse-textured or otherwise has properties indicative of high or very high saturated hydraulic conductivity throughout the mass. A satiated condition does not necessarily indicate reducing conditions. Air may be present in the water and/or the microbiological activity may be low. The presence of reducing conditions may be inferred from soil colour in some instances and a test may be performed for ferrous iron in solution. The results of the test for ferrous iron should be reported separately from the water-state description.

Evaluation: *Wet* is indicated by the occurrence of prominent water films on surfaces of sand grains and structural units that cause the soil material to glisten. If free water is absent, the term *nonsatiated wet* is used. If free water is present, the term *satiated wet* is used. The position of the upper field boundary of the satiated wet class, in a formal sense, is the top of the water in an unlined bore hole after equilibrium has been reached. Determination of the thickness of a perched zone of free water requires the installation of lined bore holes or piezometers to several depths across the zone of free water occurrence. Piezometers are tubes placed to the designated depth that are open at both ends, may have a perforated zone at the bottom, but do not permit water entry along most of their length. In the context here, information about the depth of free water and location and thickness of the free water zone would be obtained in the course of soil examination for a range of purposes and does not necessarily require installation of bore holes.

Ideally, evaluation within the moist and dry classes should be based on field instrumentation. Usually, such instrumentation is not available and approximations must be made. Gravimetric water content measurements may be used. To make the conversion from measured water content to suction it is necessary to have information on the gravimetric water retention at different suctions. The water retention at 1500 kPa may be estimated from the field clay percentage evaluation if dispersion of clay is relatively complete for the soils of concern. Commonly, the 1500 kPa retention is roughly 0.4 times the clay percentage. This relationship can be refined considerably as the soil material composition and organization is increasingly specified. Another rule of thumb is that the water content at air-dryness is about 10 percent of the clay percentage, assuming complete dispersion. Model-based curves that relate gravimetric water content and suction are available for many soils.

These curves may be used to determine upper water retention and the midpoint water retention, and to place the soil material in a water state class based on gravimetric water contents. Further such curves would be the basis in many instances for estimation of the water retention at 10 kPa from measurements at 33 kPa. A model-based curve for a medium-textured horizon and the relationship of water-state class limits to water contents determined from the desorption curve. The figure includes the results of a set of tests designed to provide local criteria for field and field office evaluation of water state. These will be discussed subsequently.

Commonly, gravimetric water content information is not available. Visual and tactile observations must suffice for the placement. Separation between moist and wet and the distinction between the two subclasses of wet may be made visually, based on water-film expression and presence of free water. Similarly, the separation between very dry and moderately dry can be made by visual or tactile comparison of the soil material at the field water content and after air drying. The change on air drying should be quite small, if the soil material initially is in the very dry class.

Criteria are more difficult to formulate for soil material that is between the moist/wet and the moderately dry/very dry separations. Four tests follow that may be useful for mineral soils. The three tests that involve tactile examination are performed on soil material that has been manipulated and mixed. This manipulation and mixing may change the tactile qualities from that of weakly altered soil material. The change may be particularly large for dense soil. In the field, this limitation should be kept in mind.

Colour Value Test: The crushed colour value of the soil for an unspecified water state is compared to the colour value at air dryness and while moderately moist or very moist. This test probably has usefulness only if the full range of colour value from air dry to moderately moist exceeds one unit of colour value. The change in colour value and its interpretation depends on the water desorption characteristics of the soil material. For example, as the water retention at 1500 kPa increases, the difference between the minimum colour value in the dry state and the very moist colour value tends to decrease.

Ball Test: A quantity of soil is squeezed firmly in the palm of the hand to form a ball about 3 to 4 cm in diameter. This is done in about five squeezes. The sphere should be near the maximum density that can be obtained by squeezing. Preparation of the ball will differ among people. The important point is that the procedure is consistent for an individual. In one approach, the ball is dropped from progressively increasing heights onto a nonresilient surface. The height in centimetres at which rupture occurs is recorded. Usually heights above 100 cm are not measured. Additionally, the manner of rupture is recorded. If the ball flattens and does not rupture, the term "deforms" is used. If the ball breaks into about five or less units, the term "pieces" is used. Finally, if the number of units exceeds about five, the term "crumbles" is used.

Alternatively, penetration resistance may be used. The penetrometer is inserted in the ball in the same fashion as would be done for soil in place. This alternative is only applicable for medium and fine- textured soil materials at higher water contents because these soil materials are relatively plastic and not subject to cracking.

Rod test. The soil material is rolled between thumb and first finger or on a surface to form a rod 3 mm in diameter or less. This rod must remain intact while being held vertically from an end for recognition as a rod. Minimum length required is 2 cm. If the maximum length that can be formed is 2 to 5 cm, the rod is weak. If the maximum length equals or exceeds 5 cm, the rod is strong.

The rod test has close similarities to the plastic limit test. Plastic limit values exceed the 1500 kPa retention at moderate clay contents and approach but are not commonly lower than the 1500 kPa retention at high clay contents. If a strong rod can be formed, the water content usually exceeds the 1500 kPa retention. The same is probably true for a weak rod. An adjustment is necessary if material of 2 to 0.5 mm is present because the plastic limit is measured on material that passes a number 40 sieve (0.43 mm in diameter).

Ribbon Test: The soil material is smeared out between thumb and first finger to form a flattened body about 2 mm of thickness. The minimum length of a coherent unit required for recognition of a ribbon is 2 cm. If the maximum length is 2 to 4 cm, the ribbon is weak. If the maximum length equals or exceeds 4 cm, the ribbon is strong.

To establish criteria based on the foregoing tests it is highly desirable to apply the tests first to soil materials that are known to be at water-state class limits. The approach would parallel that used to maintain quality control of field texture evaluation. The first step to obtain such samples is to establish gravimetric water contents for the class limits. Soil material is prepared at these water contents. A known weight of soil material at a measured, initially higher, water content than the desired final content is placed in a commercial, nylon oven-cooking bag.

These bags pass from 1 to 10 grams per hour of water at room temperature, depending on the size, the air temperature, humidity, and movement. Water loss from the bag is continued until the predetermined weight (hence, desired water content) is reached. If long-term storage is desired, the soil is next transferred to glass canning jars. The soil material either may be dried from an initially higher field water content after passing through a number 4 sieve (4.8 mm) or may be air-dried, ground, wetted to above the desired final water content, and then dried. It is preferable to pass the soil through a number 4 sieve (4.8 mm) rather than a number 10 (2 mm). The natural organization is retained to a greater extent. As a result, the calibration sample feels more like it would under field conditions. For the higher sections, consideration should be given to storage of the soil material for a day or two after the water content reduction to improve equilibration.

General relationships of the tests to water state, with the exception of the relationship of the rod test to 1500 kPa retention, have not been formulated and are probably not feasible. The tests may be applied to groupings of soils based on composition, and then locally applicable field criteria can be formulated.

Natural Drainage Classes

Natural drainage class refers to the frequency and duration of wet periods under conditions similar to those under which the soil developed. Alteration of the water regime by man, either through drainage or irrigation, is not a consideration unless the alterations have significantly changed the morphology of the soil. The classes follow:

> *Excessively drained. Water is removed very rapidly. The occurrence of internal free water commonly is very rare or very deep. The soils are commonly coarse-textured and have very high hydraulic conductivity or are very shallow.*

Somewhat excessively drained. Water is removed from the soil rapidly. Internal free water occurrence commonly is very rare or very deep. The soils are commonly coarse-textured and have high saturated hydraulic conductivity or are very shallow.

Well Drained: Water is removed from the soil readily but not rapidly. Internal free water occurrence commonly is deep or very deep; annual duration is not specified. Water is available to plants throughout most of the growing season in humid regions. Wetness does not inhibit growth of roots for significant periods during most growing seasons. The soils are mainly free of the deep to redoximorphic features that are related to wetness.

Moderately Well Drained: Water is removed from the soil somewhat slowly during some periods of the year. Internal free water occurrence commonly is moderately deep and transitory through permanent. The soils are wet for only a short time within the rooting depth during the growing season, but long enough that most mesophytic crops are affected. They commonly have a moderately low or lower saturated hydraulic conductivity in a layer within the upper 1 m, periodically receive high rainfall, or both.

Somewhat Poorly Drained: Water is removed slowly so that the soil is wet at a shallow depth for significant periods during the growing season. The occurrence of internal free water commonly is shallow to moderately deep and transitory to permanent. Wetness markedly restricts the growth of mesophytic crops, unless artificial drainage is provided. The soils commonly have one or more of the following characteristics: low or very low saturated hydraulic conductivity, a high water table, additional water from seepage, or nearly continuous rainfall.

Poorly Drained: Water is removed so slowly that the soil is wet at shallow depths periodically during the growing season or remains wet for long periods. The occurrence of internal free water is shallow or very shallow and common or persistent. Free water is commonly at or near the surface long enough during the growing season so that most mesophytic crops cannot be grown, unless the soil is artificially drained. The soil, however, is not continuously wet directly below plow-depth. Free water at shallow depth is usually present. This water table is commonly the result of low or very low saturated hydraulic conductivity of nearly continuous rainfall, or of a combination of these.

Very Poorly Drained: Water is removed from the soil so slowly that free water remains at or very near the ground surface during much of the growing season. The occurrence of internal free water is very shallow and persistent or permanent. Unless the soil is artificially drained, most mesophytic crops cannot be grown. The soils are commonly level or depressed and frequently ponded. If rainfall is high or nearly continuous, slope gradients may be greater.

Infiltration

Infiltration is the process of downward water entry into the soil. The values are usually sensitive to near surface conditions as well as to the antecedent water state. Hence, they are subject to significant change with soil use and management and time.

Infiltration stages: Three stages of infiltration may be recognized—preponded, transient ponded, and steady ponded. *Preponded infiltration* pertains to downward water entry into the soil under conditions that free water is absent on the land surface. The rate of water addition determines the rate of water entry. If rainfall intensity increases twofold, then the infiltration increases twofold. In this stage, surface-connected macropores are relatively ineffective in transporting water downward. No runoff occurs during this stage.

As water addition continues, the point may be reached where free water occurs on the ground surface. This condition is called ponding. The term in this context is less restrictive than its use in inundation. The free water may be restricted to depressions and be absent from the majority of the ground surface. Once ponding has taken place, the control over the infiltration shifts from the rate of water addition to characteristics of the soil. Surface-connected nonmatrix and subsurface-initiated cracks then become effective in transporting water downward.

Infiltration under conditions where free water is present on the ground surface is referred to as *ponded infiltration*. In the initial stages of ponded infiltration, the rate of water entry usually decreases appreciably with time because of the deeper wetting of the soil, which results in a reduced suction gradient, and the closing of cracks and other surface-connected macropores. *Transient ponded infiltration* is the stage at which the ponded infiltration decreases markedly with time. After long continued wetting under ponded conditions, the rate of infiltration becomes steady. This stage is referred to as *steady ponded infiltration*. Surface-connected cracks would be closed, if reversible. The suction gradient would be small and the driving force reduced to near that of the gravitational gradient. Assuming the absence of ice and of zones of free water within moderate depths and that surface or near surface features (crust, for example) do not control infiltration, the minimum saturated hydraulic conductivity within a depth of 1/2 to 1 meter should be a useful predictor of steady ponded infiltration rate.

Minimum Annual Steady Ponded Infiltration: The steady ponded infiltration rate while the soil is in the wettest state that regularly occurs while not frozen is called the *minimum annual steady ponded infiltration rate*. The quantity is subject to reduction because of the presence of free water at shallow depths if this is a predictable feature of the soil. Allowance for the effect of free water differentiates the quantity from minimum saturated hydraulic conductivity for the upper meter of the soil. The minimum annual steady ponded infiltration rate has application for prediction of runoff at the wettest times of the year when the runoff potential should be the highest.

Soil Temperature

Soil temperature exerts a strong influence on biological activities. It also influences the rates of chemical and physical processes within the soil. When the soil is frozen, biological activities and chemical processes essentially stop. Physical processes that are associated with ice formation are active if unfrozen zones are associated with freezing zones. Below a soil temperature of about 5 °C, growth of roots of most plants is negligible. In areas where soils have permanently frozen layers near the surface, however, even large roots of adapted plants are present immediately above the frozen layer late in the summer. Most plants grow best within a restricted range of soil and air temperature. Knowledge of soil and air temperature is essential in understanding soil-plant relationships. Temperature changes with time, as does the soil-water state. It generally differs from layer to layer at any given time.

Characteristics of Soil Temperature

Heat is both absorbed at and lost from the surface of the soil. Temperature at the surface can change in daily cycles. The soil transmits heat downward when the temperature near the surface is higher than the temperature below and heat upward when the temperature is warmer within the soil than at the surface. Soil temperatures at various depths within the soil follow cycles. The cycles deeper in the soil lag behind those near the surface. The daily cycles decrease in amplitude as depth increases and are scarcely measurable below 50 cm in most soils. Seasonal cycles are evident to much greater depths if seasonal air temperature differences are pronounced, but the temperature at a depth of 10 m is nearly constant in most soils and is about the same as the mean annual temperature of the soil above.

Soil temperature varies from layer to layer at a given site at a given time; yet, if the average annual temperatures at different depths in the same pedon are compared, they usually do not differ. Mean annual temperature is one of several useful values that describe the temperature regime of a soil.

The seasonal fluctuation of soil temperature is a characteristic of a soil. Soil temperature fluctuates little seasonally near the equator; it fluctuates widely as seasons change in the middle and high latitudes. Mean seasonal temperatures can be used to characterize soil temperature. Seasonal temperature differences decrease and the seasonal cycles lag progressively as depth increases.

For soils that freeze in winter, soil temperature is influenced by the release of heat when water changes from the liquid to the solid form. This releases about 80 calories per gram of water. The heat must be dissipated before the water in soil freezes. The rate of thaw of frozen soils is slower, because heat is required to warm the soil in order to melt the ice. In areas of heavy snowfall, the snow provides an insulating blanket and soils do not freeze as deeply or may not freeze at all.

Many factors influence soil temperature. They include amount, intensity, and distribution of precipitation; daily and monthly fluctuations in air temperature; insolation; kinds, amounts, and persistence of vegetation; duration of moisture states and snow cover; kinds of organic deposits; soil colour; aspect and gradient of slope;

elevation; and ground water. All of these factors may be described in a soil survey if they are significant.

Estimating Soil Temperature

Mean annual soil temperature in temperate, humid, continental climates can be approximated by adding 1 °C to the mean annual air temperature reported by standard meteorological stations at locations representative of the soil to be characterized. The mean annual soil temperature at a given place can be estimated more reliably by a single reading at a depth of 10 m. If water in wells is at depths between 10 and 20 m, the temperature of the water usually gives a close estimate of mean annual soil temperature. Mean annual soil temperature can also be estimated closely from the average of four readings at about 50 cm or greater depth, equally spaced throughout the year.

The mean soil temperature for summer can be estimated by averaging three measurements taken at a constant depth between 50 cm and 1 m on the 15th of each of the three months of the season. Similar methods may be used to estimate soil temperature for other seasons. These methods give values subject to minor variation caused by differences in vegetation (particularly density of canopy), ground water, snow, aspect, rain, unusual weather conditions, and other factors. Tests for nearly level, freely drained soils, both grass-covered and cultivated, produce comparable values. Over the usual period of a soil survey, systematic studies can be made to establish temperature relationships in the survey area.

Soils vary widely in the degree to which horizons are expressed. Relatively fresh geologic formations, such as fresh alluvium, sand dunes, or blankets of volcanic ash, may have no recognizable genetic horizons, although they may have distinct layers that reflect different modes of deposition. As soil formation proceeds, horizons may be detected in their early stages only by very careful examination. As age increases, horizons generally are more easily identified in the field. Only one or two different horizons may be readily apparent in some very old, deeply weathered soils in tropical areas where annual precipitation is high.

Layers of different kinds are identified by symbols. Designations are provided for layers that have been changed by soil formation and for those that have not. Each horizon designation indicates either that the original material has been changed in certain ways or that there has been little or no change. The designation is assigned after comparison of the observed properties of the layer with properties inferred for the material before it was affected by soil formation. The processes that have caused the change need not be known; properties of soils relative to those of an estimated parent material are the criteria for judgment. The parent material inferred for the horizon in question, not the material below the solum, is used as the basis of comparison. The inferred parent material commonly is very similar to, or the same as, the soil material below the solum. Designations show the investigator's interpretations of genetic relationships among the layers within a soil. Layers need not be identified by symbols for a good description; yet, the usefulness of soil descriptions is greatly enhanced by the proper use of designations.

Designations are not substitutes for descriptions. If both designations and adequate descriptions of a soil are provided, the reader has the interpretation made by the person who described the soil and also the evidence on which the interpretation was based.

Genetic horizons are not equivalent to the diagnostic horizons of Soil Taxonomy. Designations of genetic horizons express a qualitative judgment about the kind of changes that are believed to have taken place. Diagnostic horizons are quantitatively defined features used to differentiate among taxa. Changes implied by genetic horizon designations may not be large enough to justify recognition of diagnostic criteria. For example, a designation of Bt does not always indicate an argillic horizon. Furthermore, the diagnostic horizons may not be coextensive with genetic horizons.

Three kinds of symbols are used in various combinations to designate horizons and layers. These are capital letters, lower case letters, and Arabic numerals. Capital letters are used to designate the master horizons and layers; lower case letters are used as suffixes to indicate specific characteristics of master horizons and layers; and Arabic numerals are used both as suffixes to indicate vertical subdivisions within a horizon or layer and as prefixes to indicate discontinuities.

Designations for Horizons and other Layers

Master Horizons and Layers

The capital letters O, A, E, B, C, and R represent the master horizons and layers of soils. The capital letters are the base symbols to which other characters are added to complete the designations. Most horizons and layers are given a single capital letter symbol; some require two.

O horizons or layers: Layers dominated by organic material. Some are saturated with water for long periods or were once saturated but are now artificially drained; others have never been saturated.

Some O layers consist of undecomposed or partially decomposed litter, such as leaves, needles, twigs, moss, and lichens, that has been deposited on the surface; they may be on top of either mineral or organic soils. Other O layers, are organic materials that were deposited under saturated conditions and have decomposed to varying stages.

The mineral fraction of such material is only a small percentage of the volume of the material and generally is much less than half of the weight. Some soils consist entirely of material designated as O horizons or layers.

An O layer may be on the surface of a mineral soil or at any depth beneath the surface, if it is buried. A horizon formed by illuviation of organic material into a mineral subsoil is not an O horizon, although some horizons that formed in this manner contain much organic matter.

A horizons: Mineral horizons that formed at the surface or below an O horizon, that exhibit obliteration of all or much of the original rock structure, and that show one or more of the following: an accumulation of humified organic matter intimately mixed with the mineral fraction and not dominated by properties characteristic of E or B horizons (defined below) or properties resulting from cultivation, pasturing, or similar kinds of disturbance.

If a surface horizon has properties of both A and E horizons but the feature emphasized is an accumulation of humified organic matter, it is designated an A horizon. In some places, as in warm arid climates, the undisturbed surface horizon is less dark than the adjacent underlying horizon and contains only small amounts of organic matter. It has a morphology distinct from the C layer, although the mineral fraction is unaltered or only slightly altered by weathering. Such a horizon is designated A because it is at the surface; however, recent alluvial or eolian deposits that retain rock structure are not considered to be an A horizon unless cultivated.

E horizons: Mineral horizons in which the main feature is loss of silicate clay, iron, aluminium, or some combination of these, leaving a concentration of sand and silt particles. These horizons exhibit obliteration of all or much of the original rock structure.

An E horizon is usually, but not necessarily, lighter in colour than an underlying B horizon. In some soils the colour is that of the sand

and silt particles, but in many soils coatings of iron oxides or other compounds mask the colour of the primary particles. An E horizon is most commonly differentiated from an overlying A horizon by its lighter colour. It generally has less organic matter than the A horizon. An E horizon is most commonly differentiated from an underlying B horizon in the same sequum by colour of higher value, by lower chroma or both, by coarser texture, or by a combination of these properties. An E horizon is commonly near the surface below an O or A horizon and above a B horizon, but the symbol E can be used for eluvial horizons within or between parts of the B horizon or for those that extend to depths greater than normal observation if the horizon has resulted from soil genesis.

B Horizons: Horizons that formed below an A, E,, or O horizon and are dominated by obliteration of all or much of the original rock structure and show one or more of the following:

1. illuvial concentration of silicate clay, iron, aluminium, humus, carbonates, gypsum, or silica, alone or in combination;

2. evidence of removal of carbonates;

3. residual concentration of sesquioxides;

4. coatings of sesquioxides that make the horizon conspicuously lower in value, higher in chroma, or redder in hue than overlying and underlying horizons without apparent illuviation of iron;

5. alteration that forms silicate clay or liberates oxides or both and that forms granular, blocky, or prismatic structure if volume changes accompany changes in moisture content; or

6. brittleness.

All kinds of B horizons are subsurface horizons or were originally. Included as B horizons where contiguous to another genetic horizon are layers of illuvial concentration of carbonates, gypsum, or silica that are the result of pedogenic processes (these layers may or may not be cemented) and brittle layers that have other evidence of alteration, such as prismatic structure or illuvial accumulation of clay.

Examples that are not B horizons are layers in which clay films coat rock fragments or are on finely stratified unconsolidated sediments, whether the films were formed in place or by illuviation, layers into which carbonates have been illuviated but are not contiguous to an overlying genetic horizon, and layers with gleying but no other pedogenic changes.

C horizons or layers: Horizons or layers, excluding hard bedrock, that are little affected by pedogenic processes and lack properties of O, A, E, or B horizons. The material of C layers may be either like or unlike that from which the solum presumably formed. The C horizon may have been modified even if there is no evidence of pedogenesis.

Included as C layers are sediment, saprolite, unconsolidated bedrock, and other geologic materials that commonly are uncemented and exhibit low or moderate excavation difficulty. Some soils form in material that is already highly weathered. If such material does not meet the requirements of A, E, or B horizons, it is designated C. Changes not considered pedogenic are those not related to overlying horizons. Layers that have accumulations of silica, carbonates, or gypsum or more soluble salts are included in C horizons, even if indurated. If the indurated layers are obviously affected by pedogenic processes, they are a B horizon.

R layers: Hard Bedrock

Granite, basalt, quartzite and indurated limestone or sandstone are examples of bedrock that are designated R. These layers are cemented and excavation difficulty exceeds moderate. The R layer is sufficiently coherent when moist to make hand digging with a spade impractical, although it may be chipped or scraped. Some R layers can be ripped with heavy power equipment. The bedrock may contain cracks that generally are too few and too small to allow roots to penetrate at intervals of less than 10 cm. The cracks may be coated or filled with clay or other material.

Transitional and Combination Horizons

Horizons dominated by properties of one master horizon but having subordinate properties of another: Two capital letter symbols are used, as AB, EB, BE, or BC. The master horizon symbol that is given first designates the kind of horizon whose properties dominate the transitional horizon. An AB horizon, for example, has characteristics of both an overlying A horizon and an underlying B horizon, but it is more like the A than like the B. In some cases, a horizon can be designated as transitional even if one of the master horizons to which it is apparently transitional is not present. A BE horizon may be recognized in a truncated soil if its properties are similar to those of a BE horizon in a soil in which the overlying E horizon has not been removed by erosion. A BC horizon may be recognized even if no underlying C horizon is present; it is transitional to assumed parent material.

Horizons in which distinct parts have recognizable properties of the two kinds of master horizons indicated by the capital letters: The two capital letters are separated by a virgule (/), as E/B, B/E, or B/C. Most of the individual parts of one of the components are surrounded by the other.

The designation may be used even though horizons similar to one or both of the components are not present, if the separate components can be recognized. The first symbol is that of the horizon that makes up the greater volume.

Single sets of designators do not cover all situations; therefore, some improvising may be necessary. For example, Alfic Udipsamments have lamellae that are separated from each other by eluvial layers. Because it is generally not practical to describe each lamellae and eluvial layer as a separate horizon, the horizons are combined but the components are described separately.

One horizon would then contain several lamellae and eluvial layers and might be designated as an E and Bt horizon. The complete horizon sequence for this soil could be: Ap-Bw-E and Bt1-E and Bt2-C. r material.

Subordinate Distinctions Within Master Horizons and Layers

Lower case letters are used as suffixes to designate specific kinds of master horizons and layers. The word "accumulation" is used in many of the definitions in the sense that the horizon must have more of the material in question than is presumed to have been present in the parent material. The symbols and their meanings are as follows:

Highly Decomposed Organic Material

This symbol is used with "O" to indicate the most highly decomposed of the organic materials. The rubbed fibre content is less than about 17 percent of the volume.

Buried Genetic Horizon

This symbol is used in mineral soils to indicate identifiable buried horizons with major genetic features that were formed before burial. Genetic horizons may or may not have formed in the overlying material, which may be either like or unlike the assumed parent material of the buried soil. The symbol is not used in organic soils or to separate an organic layer from a mineral layer.

Concretions or Nodules

This symbol is used to indicate a significant accumulation of concretions or of nodules. Cementation is required. The cementing agent is not specified except it cannot be silica. This symbol is not used if concretions or nodules are dolomite or calcite or more soluble salts, but it is used if the nodules or concretions are enriched in minerals that contain iron, aluminium, manganese, or titanium.

Physical Root Restriction

This symbol is used to indicate root restricting layers in naturally occurring or manmade unconsolidated sediments or materials such as dense basal till, plow pans, and other mechanically compacted zones.

Organic Material of Intermediate Decomposition

This symbol is used with "O" to indicate organic materials of intermediate decomposition. Rubbed fibre content is 17 to 40 percent of the volume.

Frozen Soil

This symbol is used to indicate that the horizon or layer contains permanent ice. Symbol is not used for seasonally frozen layers or for "dry permafrost" (material that is colder than O° C but does not contain ice).

Strong Gleying

This symbol is used to indicate either that iron has been reduced and removed during soil formation or that saturation with stagnant water has preserved a reduced state. Most of the affected layers have chroma of 2 or less and many have redox concentrations. The low chroma can be the colour of reduced iron or the colour of uncoated sand and silt particles from which iron has been removed. Symbol "g" is not used for soil materials of low chroma, such as some shales or E horizons, unless they have a history of wetness. If "g" is used with "B," pedogenic change in addition to gleying is implied. If no other pedogenic change in addition to gleying has taken place, the horizon is designated Cg.

Illuvial Accumulation of Organic Matter

This symbol used with "B" to indicate the accumulation of illuvial, amorphous, dispersible organic matter-sesquioxides complexes. The sesquioxide component coats sand and silt particles. In some horizons, coatings have coalesced, filled pores, and cemented the horizon. The symbol "h" is also used in combination with "s" as "Bhs" if the amount of sesquioxide component is significant but value and chroma of the horizon are 3 or less.

Slightly Decomposed Organic Material

This symbol is used with "O" to indicate the least decomposed of the organic materials. Rubbed fibre content is more than about 40 percent of the volume.

Accumulation of Carbonates

This symbol is used to indicate the accumulation of alkaline earth carbonates, commonly calcium carbonate.

Cementation or Induration

This symbol is used to indicate continuous or nearly continuous cementation. The symbol is used only for horizons that are more than 90 percent cemented, although they may be fractured. The layer is physically root restrictive. The single predominant or codominant cementing agent may be indicated by using defined letter suffixes, singly or in pairs. If the horizon is cemented by carbonates, "km" is used; by silica, "qm"; by iron, "sm"; by gypsum, "ym"; by both lime and silica, "kqm"; by salts more soluble than gypsum, "zm."

Accumulation of Sodium

This symbol is used to indicate an accumulation of exchangeable sodium.

Residual Accumulation of Sesquioxides

This symbol is used to indicate residual accumulation of sesquioxides.

Tillage or other Disturbance

This symbol is used to indicate a disturbance of the surface layer by mechanical means, pasturing, or similar uses. A disturbed organic horizon is designated Op. A disturbed mineral horizon is designated Ap even though clearly once an E, B, or C horizon.

Accumulation of Silica

This symbol is used to indicate an accumulation of secondary silica.

Weathered or Soft Bedrock

This symbol is used with "C" to indicate root restrictive layers of soft bedrock or saprolite, such as weathered igneous rock; partly consolidated soft sandstone; siltstone; and shale. Excavation difficulty is low or moderate.

Illuvial Accumulation of Sesquioxides and Organic Matter

This symbol is used with "B" to indicate the accumulation of illuvial, amorphous, dispersible organic matter-sesquioxide complexes if both the organic matter and sesquioxide components are significant and the value and chroma of the horizon is more than 3. The symbol is also used in combination with "h" as "Bhs" if both the organic matter and sesquioxide components are significant and the value and chroma are 3 or less.

Presence of Slickensides

This symbol is used to indicate the presence of slickensides. Slickensides result directly from the swelling of clay minerals and shear failure, commonly at angles of 20 to 60 degrees above horizontal. They are indicators that other vertic characteristics, such as wedge-shaped peds and surface cracks, may be present.

Accumulation of Silicate Clay

This symbol is used to indicate an accumulation of silicate clay that has formed and subsequently translocated within the horizon or has been moved into the horizon by illuviation, or both. At least some part should show evidence of clay accumulation in the form of coatings on surfaces of peds or in pores, or as lamellae, or bridges between mineral grains.

Plinthite

This symbol is used to indicate the presence of iron-rich, humus-poor, reddish material that is firm or very firm when moist and that hardens irreversibly when exposed to the atmosphere and to repeated wetting and drying.

Development of Colour or Structure

This symbol is used with "B" to indicate the development of colour or structure, or both, with little or no apparent illuvial accumulation of material. It should not be used to indicate a transitional horizon.

Fragipan Character

This symbol is used to indicate genetically developed layers that have a combination of firmness, brittleness, very coarse prisms with few to many bleached vertical faces, and commonly higher bulk density than adjacent layers. Some part is physically root restrictive.

Accumulation of Gypsum

This symbol is used to indicate the accumulation of gypsum.

Accumulation of Salts more Soluble than Gypsum

This symbol is used to indicate an accumulation of salts more soluble than gypsum.

Conventions for using letter suffixes.—Many master horizons and layers that are symbolized by a single capital letter will have one or more lower case letter suffixes. The following rules apply:

> *Letter suffixes should immediately follow the capital letter.*

> *More than three suffixes are rarely used.*

When more than one suffix is needed, the following letters, if used, are written first: a, e, h, i, r, s, t, and w. Except for the Bhs or Crt horizons, none of these letters are used in combination in a single horizon.

If more than one suffix is needed and the horizon is not buried, these symbols, if used, are written last: c, d, f, g, m, v, and x. Some examples: Btg, Bkm, and Bsm. If a horizon is buried, the suffix "b" is written last. Suffix "b" is used only for buried mineral soils.

A B horizon that has significant accumulation of clay and also shows evidence of development of colour or structure, or both, is designated Bt ("t" has precedence over "w," "s," and "h"). A B horizon that is gleyed or that has accumulations of carbonates, sodium, silica, gypsum, salts more soluble than gypsum, or residual accumulation or sesquioxides carries the appropriate symbol—g, k, n, q, y, z, or o. If illuvial clay is also present, "t" precedes the other symbol: Btg.

Suffixes "h," "s," and "w" are not normally used with g, k, n, q, y, z, or o.

Vertical subdivision: Commonly a horizon or layer designated by a single letter or a combination of letters needs to be subdivided. The Arabic numerals used for this purpose always follow all letters. Within a C, for example, successive layers could be C1, C2, C3, and so on; or, if the lower part is gleyed and the upper part is not, the designations could be C1-C2-Cg1-Cg2 or C-Cg1-Cg2-R.

These conventions apply whatever the purpose of subdivision. In many soils, horizons that would be identified by one unique set of letters are subdivided on the basis of evident morphological features, such as structure, colour, or texture. These divisions are numbered consecutively. The numbering starts with 1 at whatever level in the profile any element of the letter symbol changes. Thus Bt1-Bt2-Btk1-Btk2 is used, not Bt1-Bt2-Btk3-Btk4. The numbering of vertical subdivisions within a horizon is not interrupted at a discontinuity (indicated by a numerical prefix) if the same letter combination is used in both materials: Bs1-Bs2-2Bs3-2Bs4 is used, not Bs1-Bs2-2Bs1-2Bs2.

Sometimes, thick layers are subdivided during sampling for laboratory analyses even though differences in morphology are not evident in the field. These layers need to be identified. This is done by following the convention of using Arabic numerals to identify the subdivision. The Arabic numerals would follow the letter designations and be a part of the horizon designation. For example, four layers of a Bt2 horizon sampled by 10-cm increments would be designated Bt21, Bt22, Bt23, and Bt24. The Bt2 horizon is subdivided for sampling purposes only.

Discontinuities: In mineral soils Arabic numerals are used as prefixes to indicate discontinuities. Wherever needed, they are used preceding A, E, B, C, and R. These prefixes are distinct from Arabic numerals used as suffixes to denote vertical subdivisions.

A discontinuity is a significant change in particle-size distribution or mineralogy that indicates a difference in the material from which the horizons formed and/or a significant difference in age, unless that difference in age is indicated by the suffix "b." Symbols to identify discontinuities are used only when they will contribute substantially to the reader's understanding of relationships among horizons. Stratification common to soils formed in alluvium is not designated as discontinuity, unless particle size distribution differs markedly (strongly contrasting particle-size class, as defined by Soil Taxonomy) from layer to layer even though genetic horizons have formed in the contrasting layers.

Where a soil has formed entirely in one kind of material, a prefix is omitted from the symbol; the whole profile is material 1. Similarly, the uppermost material in a profile having two or more contrasting materials is understood to be material 1, but the number is omitted. Numbering starts with the second layer of contrasting material, which is designated "2." Underlying contrasting layers are numbered consecutively. Even though a layer below material 2 is similar to material 1, it is designated "3" in the sequence. The numbers indicate a change in the material, not the type of material. Where two or more consecutive horizons formed in one kind of material, the same prefix number is applied to all of the horizon designations in that material: Ap-E-Bt1-2Bt2-2Bt3-2BC. The number of suffixes designating subdivisions of the Bt horizon continue in consecutive order across the discontinuity.

If an R layer is below a soil that formed in residuum and the material of the R layer is judged to be like that from which the material of the soil weathered, the Arabic number prefix is not used. If it is thought that the R layer would not produce material like that in the solum, the number prefix is used, as in A-Bt-C-2R or A-Bt-2R. If part of the solum formed in residuum, "R" is given the appropriate prefix: Ap-Bt1-2Bt2-2Bt3-2C1-2C2-2R.

Buried horizons (designated "b") are special problems. A buried horizon is obviously not in the same deposit as horizons in the overlying deposit. Some buried horizons, however, formed in material lithologically like that of the overlying deposit. A prefix is not used to distinguish material of such buried horizons. If the material in which a horizon of a buried soil formed is lithologically unlike that of the overlying material, the discontinuity is designated by number prefixes and the symbol for a buried horizon is used as well: Ap-Bt1-Bt2-BC-C-2ABb-2Btb1-2Btb2-2C.

In organic soils, discontinuities between different kinds of layers are not identified. In most cases, the differences are shown by the letter suffix designations if the different layers are organic or by the master symbol if the different layers are mineral.

Use of the prime.—Identical letter and numerical designations may be appropriate for two or more horizons separated by at least one horizon or layer of a different kind in the same pedon. The sequence A-E-Bt-E-Btx-C is an example: the soil has two E horizons. To make communication easier, the prime is used with the master horizon symbol of the lower of two horizons having identical designations: A-E-Bt-E'-Btx-C. The prime is applied to the capital letter designation and any lower-case symbols follow it: B't. The prime is not used unless all letters of the designations of two different layers are identical. Rarely, three layers have identical letter symbols; a double prime can be used: E".

The same principle applies in designating layers of organic soils. The prime is used only to distinguish two or more horizons that have identical symbols: Oi-C-O'i-C' or Oi-C-Oe-C'. The prime is added to the lower C layer to differentiate it from the upper.

Sample Horizons and Sequences

The following examples illustrate some common horizon and layer sequences of important soils and the use of Arabic numerals to identify their subdivisions. The examples were selected from soil descriptions on file and modified to reflect present conventions.

Mineral soils:

Typic Hapludoll: A1-A2-Bw-BC-C

Typic Haploboroll: Ap-A-Bw-Bk-Bky1-Bky2-C

Cumulic Haploboroll: Ap-A-Bw1-Bw2-BC-Ab-Bwb1-Bwb2-2C

Typic Argialboll: Ap-A-E-Bt1-Bt2-BC-C

Typic Argiaquoll: A-AB-BA-Btg-BCg-Cg

Entic Haplorthod: Oi-Oa-E-Bs1-Bs2-BC-C

Typic Haplorthod: Ap-E-Bhs-Bs-BC-C1-C2

Typic Fragiudalf: Oi-A-E-BE-Bt1-Bt2-B/E-Btx1-Btx2-C

Typic Haploxeralf: A1-A2-A3-2Bt1-2Bt2-2Bt3-2BC-2C

Glossoboric Hapludalf: Ap-E-B/E-Bt1-Bt2-C

Typic Paleudult: A-E-Bt1-Bt2-B/E-B't1-B't2-B't3

Typic Hapludult: 0i-A1-A2-BA-Bt1-Bt2-BC-C

Arenic Plinthic Paleudult: Ap-E-Bt-Btc-Btv1-Btv2-BC-C

Typic Haplargid: A-Bt-Bk1-Bk2-C

Entic Durorthid: A-Bw-Bq-Bqm-2Ab-2Btkb-3Byb-3Bqmb-3Bqkb

Typic Dystrochrept: Ap-Bw1-Bw2-C-R

Typic Fragiochrept: Ap-Bw-E-Bx1-Bx2-C

Typic Haplaquept: Ap-AB-Bg1-Bg2-BCg-Cg

Typic Udifluvent: Ap-C-Ab-C'

Typic Haplustert: Ap-A-AC-C1-C2

Organic soils:

Typic Medisaprist: Op-Oa1-Oa2-Oa3-C

Typic Sphagnofibrist: Oi1-Oi2-Oi3-Oe

Limnic Borofibrist: Oi-C-O'i1-O'i2-C'-Oe-C'

Lithic Cryofolist: Oi-Oa-R

Cyclic and Intermittent Horizons and Layers

A profile of a soil having cyclic horizons exposes layers whose boundaries are near the surface at one point and extend deep into the soil at another. At one place the aggregate horizon thickness may be only 50 cm; two meters away, the same horizons may be more than 125 cm thick. The cycle is repeated, commonly with considerable variation in both depth and horizontal interval, but still with some degree of regularity. If the soil is visualized in three dimensions instead of two, some cyclic horizons extend downward in inverted cones. The cone of the lower horizon fits around the cone of the horizon above. Other cyclic horizons would appear wedge-shaped.

A profile of a soil having an intermittent horizon shows that the horizon extends horizontally for some distance, ends, and reappears again some distance away. A B horizon interrupted at intervals by upward extensions of bedrock into the A horizon is an example. The distance between places where the horizon is absent is commonly variable, yet it has some degree of regularity. The distances range from less than one meter to several meters.

Obviously, a soil profile at one place could be unlike a profile only a few meters away for soils with cyclic or intermittent horizons or layers. The order of the variations of these soils are given in soil descriptions.

Descriptions of the order of horizontal variation within a pedon include the kind of variation, the spacing of cycles or interruptions, and the amplitude of depth variation of cyclic horizons.

Near Surface Subzones

The morphology of the uppermost few centimetres is subject in many soils to strong control by antecedent weather and by soil use. A soil may be freshly tilled today and have a loose surface. Tomorrow it may have a strong crust because of a heavy rain. Or, in one place soil may be highly compacted by livestock and have a firm near surface even though over most of its extent the same uppermost few centimetres are little disturbed and very friable. There is a need for a set of terms to describe subzones of the near surface and, in particular, the near surface of tilled soils. Five subzones of the near surface are recognized.

The *mechanically bulked* subzone has undergone through mechanical manipulation a reduction in bulk density and an increase in discreteness of structural units, if present. Usually the mechanical manipulation is the consequence of tillage operations. Rupture resistance of the mass overall, inclusive of a number of structural units, is *loose* or *very friable* and *occasionally friable*. Individual structural units may be *friable* or even *firm*. Mechanical continuity among structural units is low. Structure grade, if the soil material exhibits structural units < 20 mm across, is moderate or strong. Strain that results from contraction on drying of individual structural units may not extend among structural units. Hence, internally initiated desiccation cracks may be weak or absent even though the soil material in a consolidated condition has considerable potential extensibility. Cracks may be present, however, if they are initiated deeper in the soil.

The *mechanically compacted* subzone has been subject to compaction, usually in tillage operations but possibly by animals. Commonly, mechanical continuity of the fabric and bulk density are increased. Rupture resistance depends on texture and degree of compaction. Generally, *friable* is the minimum class. Mechanical continuity of the fabric permits propagation of strain that results on drying only over several centimetres. Internally initiated cracks appear if the soil material has appreciable extensibility and drying has been sufficient. In some soils this subzone restricts root growth. The suffix "d" may be used if compaction results in a strong plow pan.

The *water-compacted* subzone has been compacted by repetitive large changes in water state without mechanical load except for the weight of the soil. Repetitive occurrence of free water is particularly conducive to compaction. Depending on texture, moist rupture resistance ranges from *very friable* through firm. Structural units, if present, are less discrete than for the same soil material if mechanically bulked. Structure generally would be weak or the condition would be massive. Mechanical continuity of the fabric is sufficient that strain which originates on drying propagates appreciable distances. As a consequence, if extensibility is sufficient, cracks develop on drying. In many soils, over time the water-compacted subzone replaces the mechanically bulked subzone. The replacement can occur in a single year if the subzone is subject to periodic occurrence of free water with intervening periods when *slightly moist* or *dry*. The presence of a water-compacted subzone and the absence of the mechanically bulked subzone is an important consequence of no-till farming systems.

The *surficial bulked* subzone occurs in the very near surface. Continuity of the fabric is low. Cracks are not initiated in this subzone, although they may be present if initiated in underlying more compacted soil. The subzone is formed by various processes. Frost action under conditions where the soil is drier than wet is a mechanism. Wetting and drying of soil material with high extensibility is another origin; certain Vertisols are illustrative.

Crust is a surficial subzone, usually less than 50 mm thick, that exhibits markedly more mechanical continuity of the soil fabric than the zone immediately beneath. Commonly, the original soil fabric has been reconstituted by water action and the original structure has been replaced by a massive condition. While the material is wet, raindrop impact and freeze-thaw cycles are mechanisms leading to reconstitution. Crusting related to *raindrop-impact* and *freeze-thaw* are recognized.

A *fluventic* zone may be formed by local transport and deposition of soil material in tilled fields. Such a feature has weaker mechanical continuity than a crust. The rupture resistance is lower, and the reduction in infiltration may be less than for crusts of similar texture. A raindrop-impact crust may occur on a fluventic zone.

Crusts and a fluventic zone may be described in terms of thickness in millimeters, structure and other aspects of the fabric, and by consistence, including rupture resistance while dry and micropenetration resistance while wet. Thickness pertains to the zone where reconstitution of the fabric has been pronounced. Also, the distance between *surface-initiated cracks* may be a useful observation for seedling emergence considerations. If the distance is short, the weight of the crust slabs is low. Soil material with little apparent reconstitution commonly adheres beneath the crust and is removed with the crust. This soil material that shows little or no reconstitution is not part of the crust and does not contribute to the thickness.

Identification of subzones is not clear cut. Morphological expression of bulking and compaction may be quite different among soils dependent on particle size distribution, organic matter content, clay mineralogy, water regime, and possibly other factors.

The distinction between a bulked and compacted state for soil material with appreciable extensibility is made in part on the potential for the transmission of strain on drying over distances greater than the horizontal dimensions of the larger structural units. In a bulked subzone little or no strain is propagated; in a compacted subzone the strain would be propagated over distances greater than the horizontal dimensions of the larger structural units. Many soils have low extensibility because of texture, clay mineralogy, or both. For these soils, the expression of cracks cannot be used to distinguish between a bulked and compacted state.

The distinction between compaction and bulking is subjective. It is useful to establish a concept of a normal degree of compaction of the near surface to which the actual degree of compaction is compared. The concept for tilled soils should be the compaction of soil material on level or convex parts of the tillage determined relief. The soil should have been subject to the bulking action of conventional tillage without the subsequent mechanical compaction. The subzone in question should have been brought to a *wet* or *very moist* water state from an appreciably drier condition followed by drying to *slightly moist* or drier at least once. It should not have been subject, however, to a large number of wetting and drying cycles where the maximum wetness involves the presence of free water. If the soil material has a degree of compaction similar to what would be expected, then the term *normal compaction* is employed.

Boundaries of Horizons and Layers

A boundary is a surface or transitional layer between two adjoining horizons or layers. Most boundaries are zones of transition rather than sharp lines of division. Boundaries vary in distinctness and in topography.

Distinctness: Distinctness refers to the thickness of the zone within which the boundary can be located. The distinctness of a boundary depends partly on the degree of contrast between the adjacent layers and partly on the thickness of the transitional zone between them. Distinctness is defined in terms of thickness of the transitional zone:

Abrupt: Less than 2 cm thick

Clear: 2 to 5 cm thick

Gradual: 5 to 15 cm thick

Diffuse: More than 15 cm thick

Abrupt soil boundaries, such as those between the E and Bt horizons in many soils, are easily determined. Some boundaries are not readily seen but can be located by testing the soil above and below the boundary. Diffuse boundaries, such as those in many old soils in tropical areas, are most difficult to locate and require time-consuming comparisons of small specimens of soil from various parts of the profile until the midpoint of the transitional zone is determined. For soils that have nearly uniform properties or that change very gradually as depth increases, horizon boundaries are imposed more or less arbitrarily without clear evidence of differences.

Topography: Topography refers to the irregularities of the surface that divides the horizons. Even though soil layers are commonly seen in vertical section, they are three-dimensional. Topography of boundaries is described with the following terms:

Smooth: The boundary is a plane with few or no irregularities.

Wavy: The boundary has undulations in which depressions are wider then they are deep.

Irregular: The boundary has pockets that are deeper than they are wide.

Broken: One or both of the horizons or layers separated by the boundary are discontinuous and the boundary is interrupted.

Root Restricting Depth

The root restricting depth is where root penetration would be strongly inhibited because of physical (including soil temperature) and/or chemical characteristics. Restriction means the incapability to support more than a few *fine* or *very fine* roots if depth from the soil surface and water state, other than the occurrence of frozen water, are not limiting. For cotton or soybeans and possibly other crops with less abundant roots than the grasses, the *very few* class is used instead of the *few* class. The restriction may be below where plant roots normally occur because of limitations in water state, temperatures, or depth from the surface. The evaluation should be for the specific plants that are important to the use of the soil. These plants should be indicated. The root-restriction depth may differ depending on the plant considered.

Root-depth observations preferably should be used to make the generalization. If these are not available—and often they are not because roots do not extend to the depth of concern—then inferences may be made from morphology. Some guidelines follow for physical restriction. Chemical restrictions, such as high extractable aluminium and/or low extractable calcium, are not considered here. These are generally not determinable by field examination alone.

Physical root restriction is assumed at contact to rock, whether hard or soft. Further, certain pedogenic horizons, such as *fragipans*, infer root restriction. A change in particle size distribution alone, as for example loamy sand over gravel, is not always a basis for physical root restriction.

A common indication of physical root restriction is a combination of structure and consistence which together suggest that the resistance of the soil fabric to root entry is high and that vertical cracks and planes of weakness for root entry are absent or widely spaced.

Root restriction is inferred for a continuously *cemented* zone of any thickness; or a zone >10-cm thick that when *very moist* or *wet* is *massive*, *platy*, or has *weak* structure of any type for a vertical repeat distance of >10 cm and while *very moist* or *wet* is *very firm* (*firm*, if sandy), *extremely firm*, or has a *large* penetration resistance.

Classes of root-restricting depth:

Very shallow:	< 25 cm
Shallow:	25-50 cm
Moderately deep:	50-100 cm
Deep:	100-150 cm
Very deep:	>150 cm

Particle Size Distribution

This section discusses particle distribution. The finer sizes are called *fine earth* (smaller than 2 mm diameter) as distinct from *rock fragments* (pebbles, cobbles, stones, and boulders). Particle-size distribution of fine earth or less than 2 mm is determined in the field mainly by feel. The content of rock fragments is determined by estimating the proportion of the soil volume that they occupy.

Soil Separates

The United States Department of Agriculture uses the following size separates for the <2 mm mineral material:

Name Size (mm)

Very coarse sand:	2.0-1.0 mm
Coarse sand:	1.0-0.5 mm
Medium sand:	0.5-0.25 mm
Fine sand:	0.25-0.10 mm
Very fine sand:	0.10-0.05 mm
Silt:	0.05-0.002 mm
Clay:	< 0.002 mm

Soil Texture

Soil texture refers to the weight proportion of the separates for particles less than 2 mm as determined from a laboratory particle-size distribution. Field estimates should be checked against laboratory determinations and the field criteria should be adjusted as necessary. Some soils are not dispersed completely in the standard particle size analysis. For these, the field texture is referred to as *apparent* because it is not an estimate of the results of a laboratory operation. Apparent field texture is a tactile evaluation only with no inference as to laboratory test results. Field criteria for estimating soil texture must be chosen to fit the soils of the area. Sand particles feel gritty and can be seen individually with the naked eye. Silt particles cannot be seen individually without magnification; they have a smooth feel to the fingers when dry or wet. In some places, clay soils are sticky; in others they are not. Soils dominated by montmorillonite clays, for example, feel different from soils that contain similar amounts of micaceous or kaolintic clay. Even locally, the relationships that are useful for judging texture of one kind of soil may not apply as well to another kind.

The texture classes are sand, loamy sands, sandy loams, loam, silt loam, silt, sandy clay loam, clay loam, silty clay loam, sandy clay, silty clay, and clay. Subclasses of sand are subdivided into coarse sand, sand, fine sand, and very fine sand. Subclasses of loamy sands and sandy loams that are based on sand size are named similarly.

Definitions of the soil texture classes follow:

Sands: More than 85 percent sand, the percentage of silt plus 1.5 times the percentage of clay is less than 15.

Coarse sand. A total of 25 percent or more very coarse and coarse sand and less than 50 percent any other single grade of sand.

Sand: A total of 25 percent or more very coarse, coarse, and medium sand, a total of less than 25 percent very coarse and coarse sand, and less than 50 percent fine sand and less than 50 percent very fine sand.

Fine Sand: 50 percent or more fine sand; or a total of less than 25 percent very coarse, coarse, and medium sand and less than 50 percent very fine sand.

Very fine sand. 50 percent or more very fine sand.

Loamy Sands: Between 70 and 91 percent sand and the percentage of silt plus 1.5 times the percentage of clay is 15 or more; and the percentage of silt plus twice the percentage of clay is less than 30.

Loamy Coarse Sand: A total of 25 percent or more very coarse and coarse sand and less than 50 percent any other single grade of sand.

Loamy Sand: A total of 25 percent or more very coarse, coarse, and medium sand and a total of less than 25 percent very coarse and coarse sand, and less than 50 percent fine sand and less than 50 percent very fine sand.

Loamy Fine Sand: 50 percent or more fine sand; or less than 50 percent very fine sand and a total of less than 25 percent very coarse, coarse, and medium sand.

Loamy Very Fine Sand: 50 percent or more very fine sand.

Sandy Loams: 7 to 20 percent clay, more than 52 percent sand, and the percentage of silt plus twice the percentage of clay is 30 or more; or less than 7 percent clay, less than 50 percent silt, and more than 43 percent sand.

Coarse Sandy Loam: A total of 25 percent or more very coarse and coarse sand and less than 50 percent any other single grade of sand.

Sandy Loam: A total of 30 percent or more very coarse, coarse, and medium sand, but a total of less than 25 percent very coarse and coarse sand and less than 30 percent fine sand and less than 30 percent very fine sand; or a total of 15 percent or less very coarse, coarse, and medium sand, less than 30 percent fine sand and less than 30 percent very fine sand with a total of 40 percent or less fine and very fine sand.

Fine Sandy Loam: 30 percent or more fine sand and less than 30 percent very fine sand; or a total of 15 to 30 percent very coarse, coarse, and medium sand; or a total of more than 40 percent fine and very fine sand, one half or more of which is fine sand, and a total of 15 percent or less very coarse, coarse, and medium sand.

Very Fine Sandy Loam: 30 percent or more very fine sand and a total of less than 15 percent very coarse, coarse, and medium sand; or more than 40 percent fine and very fine sand, more than one half of which is very fine sand, and total of less than 15 percent very coarse, coarse, and medium sand.

Loam: 7 to 27 percent clay, 28 to 50 percent silt, and 52 percent or less sand.

Silt Loam: 50 percent or more silt and 12 to 27 percent clay, or 50 to 80 percent silt and less than 12 percent clay.

Silt: 80 percent or more silt and less than 12 percent clay.

Sandy Clay Loam: 20 to 35 percent clay, less than 28 percent silt, and more than 45 percent sand.

Clay Loam: 27 to 40 percent clay and more than 20 to 46 percent sand.

Silty Clay Loam: 27 to 40 percent clay and 20 percent or less sand.

Sandy Clay: 35 percent or more clay and 45 percent or more sand.

Silty Clay: 40 percent or more clay and 40 percent or more silt.

Clay: 40 percent or more clay, 45 percent or less sand, and less than 40 percent silt.

The texture triangle is used to resolve problems related to word definitions, which are somewhat complicated. The eight distinctions in the sand and loamy sand groups provide refinement greater than can be consistently determined by field techniques. Only those distinctions that are significant to use and management and that can be consistently made in the field should be applied.

Groupings of Soil Texture Classes: The need for fine distinctions in the texture of the soil layers results in a large number of classes of soil texture. Often it is convenient to speak generally of broad groups or classes of texture. An outline of soil texture groups, in three classes and in five, follows. In some areas where soils are high in silt, a fourth general class, silty soils, may be used for silt and silt loam.

Organic Soils

Layers that are not saturated with water for more than a few days at a time are organic if they have 20 percent or more organic carbon. Layers that are saturated for longer periods, or were saturated before being drained, are organic if they have 12 percent or more organic carbon and no clay, 18 percent or more organic carbon and 60 percent or more clay, or a proportional amount of organic carbon, between 12 and 18 percent, if the clay content is between 0 and 60 percent.

The kind and amount of the mineral fraction, the kind of organisms from which the organic material was derived, and the state of decomposition affect the properties of the soil material. Descriptions include the percentage of undecomposed fibres and the solubility in sodium pyrophosphate of the humified material. A special effort is made to identify and estimate the volume occupied by *sphagnum* fibres, which have extraordinary high water retention characteristics. When squeezed firmly in the hand to remove as much water as possible, *sphagnum* fibres are lighter in colour than fibres of *hypnum* and most other mosses.

Fragments of wood more than 2 cm across and so undecomposed that they cannot be crushed by the fingers when moist or wet are called "wood fragments." They are comparable to rock fragments in mineral soils and are described in a comparable manner.

Muck (sapric) is well-decomposed, organic soil material. *Peat* (fibric) is relatively undecomposed, organic material in which the original fibres constitute almost all of the material. *Mucky peat* (hemic) is material intermediate between muck and peat.

Rock Fragments

Rock fragments are unattached pieces of rock 2 mm in diameter or larger that are *strongly cemented* or more resistant to rupture. Rock fragments include all sizes that have horizontal dimensions less than the size of a pedon.

Rock fragments are described by size, shape, and, for some, the kind of rock. The classes are *pebbles, cobbles, channers, flagstones, stones, and boulders*. If a size or range of sizes predominates, the class is modified, as for example: "fine pebbles," "cobbles 100 to 150 mm in diameter," "channers 25 to 50 mm in length."

Gravel is a collection of pebbles that have diameters ranging from 2 to 75 mm. The term is applied to the collection of pebbles in a soil layer with no implication of geological formalization.

The terms "pebble" and "cobble" are usually restricted to rounded or subrounded fragments; however, they can be used to describe angular fragments if they are not flat. Words like chert, limestone, and shale refer to a kind of rock, not a piece of rock. The composition of the fragments can be given: "chert pebbles," "limestone channers." The upper size of gravel is 3 inches (75 mm). This coincides with the upper limit used by many engineers for grain-size distribution computations. The 5-mm and 20-mm divisions for the separation of fine, medium, and coarse gravel coincide with the sizes of openings in the "number 4" screen (4.76 mm) and the "3/4 inch" screen (19.05 mm) used in engineering.

The 75 mm (3 inch) limit separates gravel from cobbles. The 250-mm (10-inch) limit separates cobbles from stones, and the 600-mm (24-inch) limit separates stones from boulders. The 150-mm (channers) and 380 mm (flagstones) limits for thin, flat fragments follow conventions used for many years to provide class limits for plate-shaped and crudely spherical rock fragments that have about the same soil use implications as the 250-mm limit for spherical shapes.

Rock Fragments in the Soil

Historically, the total volume of rock fragments of all sizes has been used to form classes. The interpretations program imposes requirements that cannot be met by grouping all sizes of rock fragments together. Furthermore, the interpretations program requires weight rather than volume estimates. For interpretations, the weight percent >250, 75-250, 5-75 and 2-5 mm are required; the first two are on a whole soil basis, and the latter two are on a <75 mm basis. For the >250 and 75-250 mm, weighing is generally impracticable. Volume percentage estimates would be made from areal percentage measurements by point-count or line-intersect methods.

Length of the transect or area of the exposure should be 50 and preferably 100 times the area or dimensions of the rock fragment size that encompasses about 90 percent of the rock fragment volume. For the <75 mm weight, measurements are feasible but may require 50-60 kg of sample if appreciable rock fragments near 75 mm are present. An alternative is to obtain volume estimates for the 20-75 mm and weight estimates for the <20 mm. This is favoured because of the difficulty in visual evaluation of the 2 to 5 mm size separations. The weight percentages of 5-20 and 2-5 mm may be converted to volume estimates and placed on a <75 mm base by computation. The adjectival form of a class name of rock fragments is used as a modifier of the textural class name: "gravelly loam," "stony loam."

Soil Colour

Elements of soil colour descriptions are the colour name, the Munsell notation, the water state, and the physical state: "brown (10YR 5/3), dry, crushed, and smoothed."

Physical state is recorded as broken, rubbed, crushed, or crushed and smoothed. The term "crushed" usually applies to dry samples and "rubbed" to moist samples. If unspecified, the surface is broken. The colour of the soil is recorded for a surface broken through a ped if a ped can be broken as a unit.

The colour value of most soil material becomes lower after moistening. Consequently, the water state of a sample is always given. The water state is either "moist" or "dry." The dry state for colour determinations is air-dry and should be made at the point where the colour does not change with additional drying. Colour in the moist state is determined on moderately moist or very moist soil material and should be made at the point where the colour does not change with additional moistening. The soil should not be moistened to the extent that glistening takes place as colour determinations of wet soil may be in error because of the light reflection of water films. In a humid region, the moist state generally is considered standard;

in an arid region, the dry state is standard. In detailed descriptions, colours of both dry and moist soil are recorded if feasible. The colour for the regionally standard moisture state is usually described first. Both moist and dry colours are particularly valuable for the immediate surface and tilled horizons in order to assess reflectance.

Munsell notation is obtained by comparison with a Munsell system colour chart. The most commonly used chart includes only about one fifth of the entire range of hues. It consists of about 250 different coloured papers, or chips, systematically arranged on hue cards according to their Munsell notations. The arrangements of colour chips on a Munsell colour card.

The Munsell colour system uses three elements of colour—*hue, value,* and *chroma*—to make up a colour notation. The notation is recorded in the form: hue, value/chroma—for example, 5Y 6/3.

Hue is a measure of the chromatic composition of light that reaches the eye. The Munsell system is based on five principal hues: red (R), yellow (Y), green (G), blue (B), and purple (P). Five intermediate hues representing midpoints between each pair of principal hues complete the 10 major hue names used to describe the notation. The intermediate hues are yellow-red (YR), green-yellow (GY), blue-green (BG), purple-blue (PB), and red-purple (RP). The relationships among the 10 hues. Each of the 10 major hues is divided into four segments of equal visual steps, which are designated by numerical values applied as prefixes to the symbol for the hue name. For example, 10R marks a limit of red hue. Four equally spaced steps of the adjacent yellow-red (YR) hue are identified as 2.5YR, 5YR, 7.5YR, and 10YR respectively. The standard chart for soil has separate hue cards from 10R through 5Y.

Value indicates the degree of lightness or darkness of a colour in relation to a neutral gray scale. On a neutral gray (achromatic) scale, value extends from pure black (0/) to pure white (10/). The value notation is a measure of the amount of light that reaches the eye under standard lighting conditions. Gray is perceived as about halfway between black and white and has a value notation of 5/. The actual amount of light that reaches the eye is related logarithmically to colour value. Lighter colours are indicated by numbers between 5/ and 10/; darker colours are indicated by numbers from 5/ to 0/. These values may be designated for either achromatic or chromatic conditions. Thus, a card of the colour chart for soil has a series of chips arranged vertically to show equal steps from the lightest to the darkest shades of that hue. This arrangement vertically on the card for the hue of 10YR.

Chroma is the relative purity or strength of the spectral colour. Chroma indicates the degree of saturation of neutral gray by the spectral colour. The scales of chroma for soils extend from /0 for neutral colours to a chroma of /8 as the strongest expression of colour used for soils. That colour chips are arranged horizontally by increasing chroma from left to right on the colour card.

The complete colour notation can be visualized. Pale brown, for example, is designated 10YR 6/3. Very dark brown is designated 10YR 2/2. All of the colours on the chart have hue of 10YR. The darkest shades of that hue are at the bottom of the card and the lightest shades are at the top. The weakest expression of chroma (the grayest colour) is at the left; the strongest expression of chroma is at the right.

At the extreme left of the card are symbols such as N 6/. These are colours of zero chroma which are totally achromatic—neutral colour. They have no hue and no chroma but range in value from black (N 2/) to white (N 8/). An example of a notation for a neutral (achromatic) colour is N 5/ (gray). The colour 10YR 5/1 is also called "gray," for the hue is hardly perceptible at such low chroma.

Conditions for measuring colour: The quality and intensity of the light affect the amount and quality of the light reflected from the sample to the eye. The moisture content of the sample and the roughness of its surface affect the light reflected. The visual impression of colour from the standard colour chips is accurate only under standard conditions of light intensity and quality. Colour determination may be inaccurate early in the morning or late in the evening. When the sun is low in the sky or the atmosphere is smoky, the light reaching the sample and the light reflected is redder. Even though the same kind of light reaches the colour standard and the sample, the reading of sample colour at these times is commonly one or more intervals of hue redder than at midday.

Colours also appear different in the subdued light of a cloudy day than in bright sunlight. If artificial light is used, as for colour determinations in an office, the light source used must be as near the white light of midday as possible. With practice, compensation can be made for the differences unless the light is so subdued that the distinctions between colour chips are not apparent. The intensity of incidental light is especially critical when matching soil to chips of low chroma and low value.

Roughness of the reflecting surface affects the amount of reflected light, especially if the incidental light falls at an acute angle. The incidental light should be as nearly as possible at a right angle. For crushed samples, the surface is smoothed; the state is recorded as "dry, crushed, and smoothed."

Recording Guidelines

Uncertainty: Under field conditions, measurements of colour are reproducible by different individuals within 2.5 units of hue (one card) and 1 unit of value and chroma. Notations are made to the nearest whole unit of value and chroma.

Before 1989, the cards for hues of 2.5YR, 7.5YR, and 2.5Y did not include chips for colours having chroma of 3. These colours are encountered frequently in some soils and can be estimated reliably by interpolation between adjacent chips of the same hue. Chips for chromas of 5 and 7 are not provided on any of the standard colour cards. Determinations are usually not precise enough to justify interpolation between chromas of 4 and 6 or 6 and 8. Colour should never be extrapolated beyond the highest chip. It should also be rounded to the nearest chip. For many purposes, the differences between colours of some adjacent colour chips have little significance. For such purposes, colour notations have been grouped, and the groups have been named.

Dominant Colour

The dominant colour is the colour that occupies the greatest volume of the layer. Dominant colour (or colours) is always given first among those of a multicolored layer. It is judged on the basis of colours of a broken sample. For only two colours, the dominant colour makes up more than 50 percent of the volume. For three or more colours, the dominant colour makes up more of the volume of the layer than any other colour, although it may occupy less than 50 percent. The expression "brown with yellowish brown and grayish brown" signifies that brown is the dominant colour. It may or may not make up more than 50 percent of the layer.

In some layers, no single colour is dominant and the first colour listed is not more prevalent than others. The expression "brown and yellowish brown with grayish brown" indicates that brown and yellowish brown are about equal and are codominant. If the colours are described as "brown, yellowish brown, and grayish brown," the three colours make up nearly equal parts of the layer.

Mottling

Mottling refers to repetitive colour changes that cannot be associated with compositional properties of the soil. Redoximorphic features are a type of mottling that is associated with wetness. A colour pattern that can be related to proximity to a ped surface or other organizational or compositional feature is not mottling. Mottle description follows the dominant colour. Mottles are described by quantity, size, contrast, colour, and other attributes in that order.

Quantity is indicated by three areal percentage classes of the observed surface:

few: less than 2 percent,

common: 2 to 20 percent, and

many: more than 20 percent.

The notations must clearly indicate to which colours the terms for quantity apply. For example, "common grayish brown and yellowish brown mottles" could mean that each makes up 2 to 20 percent of the horizon. By convention, the example is interpreted to mean that the quantity of the two colours *together* is between 2 and 20 percent. If each colour makes up between 2 and 20 percent, the description should read "common grayish brown (10YR 5/2) and common yellowish brown (10YR 5/4) mottles."

Size refers to dimensions as seen on a plane surface. If the length of a mottle is not more than two or three times the width, the dimension recorded is the greater of the two. If the mottle is long and narrow, as a band of colour at the periphery of a ped, the dimension recorded is the smaller of the two and the shape and location are also described. Three size classes are used:

fine: smaller than 5 mm,

medium: 5 to 15 mm, and

coarse: larger than 155 mm.

Contrast refers to the degree of visual distinction that is evident between associated colours:

Faint: Evident only on close examination. Faint mottles commonly have the same hue as the colour to which they are compared and differ by no more than 1 unit of chroma or 2 units of value. Some faint mottles of similar but low chroma and value differ by 2.5 units (one card) of hue.

Distinct: Readily seen but contrast only moderately with the colour to which they are compared. Distinct mottles commonly have the same hue as the colour to which they are compared but differ by 2 to 4 units of chroma or 3 to 4 units of value; or differ from the colour to which they are compared by 2.5 units (one card) of hue but by no more than 1 unit of chroma or 2 units of value.

Prominent: Contrast strongly with the colour to which they are compared. Prominent mottles are commonly the most obvious colour feature of the section described. Prominent mottles that have medium chroma and value commonly differ from the colour to which they are compared by at least 5 units (two pages) of hue if chroma and value are the same; at least 4 units of value or chroma if the hue is the same; or at least 1 unit of chroma or 2 units of value if hue differs by 2.5 units.

Contrast is often not a simple comparison of one colour with another but is a visual impression of the prominence of one colour against a background commonly involving several colours.

Shape, location, and character of boundaries of mottles are indicated as needed. *Shape* is described by common words such as streaks, bands, tongues, tubes, and spots. *Location* of mottles as related to structure of the soil may be significant. *Boundaries* may be described as *sharp* (colour gradation is not discernable with the naked eye), *clear* (colour grades over less than 2 mm), or *diffuse* (colour grades over more than 2 mm).

Moisture state and physical state of the dominant colour are presumed to apply to the mottles unless the description states otherwise. An example, for which a standard moist broken state of the sample has been specified, might read "brown (10YR 4/3), brown (10YR 5/3) dry; many medium distinct yellowish brown (10YR 5/6) mottles, brownish yellow (10YR 6/6) dry." Alternatively, the colours in the standard moisture state may be given together, followed by the colours at other moisture states. The colour of mottles commonly is given only for the standard state unless special significance can be attached to colours at another state.

A nearly equal mixture of two colours for a moist broken standard state can be written "intermingled brown (10YR 4/3) and yellowish brown (10YR 5/6) in a medium distinct pattern; brown (10YR 5/3) and brownish yellow (10YR 6/6) dry." If a third colour is present, "common medium faint dark grayish brown (10YR 4/2) mottles, grayish brown (10YR 5/2) dry" can be added.

If the mottles are fine and faint so that they cannot be compared easily with the colour standards, the Munsell notation should be omitted. Other abbreviated descriptions are used for specific circumstances.

Colour Patterns

Colour, including mottling, may be described separately for any feature that may merit a separate description, such as peds, concretions, nodules, cemented bodies, filled animal burrows, and the like. Colour patterns that exhibit a spatial relationship to composition changes or to features such as nodules or surfaces of structural units may be useful to record because of the inferences that may be drawn about genesis and soil behaviour. Colours may be given for extensions of material from another soil layer. The fine tubular colour patterns that extend vertically below the A horizon of some wet soils, for example, were determined by the environment adjacent to roots that once occupied the tubules. The rim of bright colour within an outer layer of lighter colour at the surface of some peds relate to water movement into and out of the peds and to oxidation-reduction relationships.

Ground surface colour: The colour value of the immediate ground surface may differ markedly from that of the surface horizon. For example, raindrop impact may have removed clay-size material from the surface of sand and silt which results in a surficial millimetre or so of increased colour value. In some arid soils, dark rock fragments may have reduced the colour value of the ground surface appreciably from that of the fine earth of the surface horizon as a whole. Furthermore, dead vegetation may have colour values that differ appreciably from those for the fine earth of the surface horizon. Colour information is, therefore, desirable for the actual ground surface inclusive of the vegetation as well as the soil material. Surface colour influences reflectivity of light, therefore, the capacity to absorb and release radiant energy.

Surface soil colours commonly range widely at a site, and it may be necessary to array mentally the colour values and their areal proportion for the ground surface, whether rock fragments, dead vegetation, or fine earth. Then a single colour value is selected for each important ground surface component. From the areal proportion of the components, and their colour value, a weighted average colour value for the ground surface may be computed. Estimation of the areal proportion of components is discussed in the section on ground cover.

Soil Structure

Soil structure refers to units composed of primary particles. The cohesion within these units is greater than the adhesion among units.

As a consequence, under stress, the soil mass tends to rupture along predetermined planes or zones. These planes or zones, in turn, form the boundary. Compositional differences of the fabric matrix appear to exert weak or no control over where the bounding surfaces occur. If compositional differences control the bounding surfaces of the body, then the term "concentration" is employed. The term "structural unit" is used for any repetitive soil body that is commonly bounded by planes or zones of weakness that are not an apparent consequence of compositional differences. A structural unit that is the consequence of soil development is called a ped.

The surfaces of peds persist through cycles of wetting and drying in place. Commonly, the surface of the ped and its interior differ as to composition or organization, or both, because of soil development. Earthy *clods* and *fragments* stand in contrast to peds, for which soil forming processes exert weak or no control on the boundaries. Some clods, adjacent to the surface of the body, exhibit some rearrangement of primary particles to a denser configuration through mechanical means. The same terms and criteria used to describe structured soils should be used to describe the shape, grade, and size of clods. Structure is not inferred by using the terms interchangeably. A size sufficient to affect tilth adversely must be considered.

The distinction between clods and fragments rests on the degree of consolidation by mechanical means. Soil fragments include (1) units of undisturbed soil with bounding planes of weakness that are formed on drying without application of external force and which do not appear to have predetermined bounding planes, (2) units of soil disturbed by mechanical means but without significant rearrangement to a denser configuration, and (3) pieces of soil bounded by planes of weakness caused by pressure exerted during examination with size and shape highly dependent on the manner of manipulation.

Some soils lack structure and are referred to as *structureless*. In structureless layers or horizons, no units are observable in place or after the soil has been gently disturbed, such as by tapping a spade containing a slice of soil against a hard surface or dropping a large fragment on the ground. When structureless soils are ruptured, soil fragments, single grains, or both result. Structureless soil material may be either single grain or massive. Soil material of single grains lacks structure. In addition, it is loose. On rupture, more than 50 percent of the mass consists of discrete mineral particles.

Some soils have *simple structure*, each unit being an entity without component smaller units. Others have *compound structure*, in which large units are composed of smaller units separated by persistent planes of weakness.

In soils that have structure, the shape, size, and grade (distinctness) of the units are described. Field terminology for soil structure consists of separate sets of terms designating each of the three properties, which by combination form the names for structure.

Shape: Several basic shapes of structural units are recognized in soils. Supplemental statements about the variations in shape of individual peds are needed in detailed descriptions of some soils. The following terms describe the basic shapes and related arrangements:

Platy: The units are flat and platelike. They are generally oriented horizontally. Platy structure is illustrate. A special form, lenticular platy structure, is recognized for plates that are thickest in the middle and thin toward the edges.

Prismatic: The individual units are bounded by flat to rounded vertical faces. Units are distinctly longer vertically, and the faces are typically casts or molds of adjoining units. Vertices are angular or subrounded; the tops of the prisms are somewhat indistinct and normally flat. Prismatic structure is illustrate.

Columnar: The units are similar to prisms and are bounded by flat or slightly rounded vertical faces. The tops of columns, in contrast to those of prisms, are very distinct and normally rounded.

Blocky: The units are blocklike or polyhedral. They are bounded by flat or slightly rounded surfaces that are casts of the faces of surrounding peds. Typically, blocky structural units are nearly equidimensional but grade to prisms and to plates. The structure is described as angular blocky if the faces intersect at relatively sharp angles; as subangular blocky if the faces are a mixture of rounded and plane faces and the corners are mostly rounded.

Size: Five classes are employed: very fine, fine, medium, coarse, and very coarse. The size limits of the classes differ according to the shape of the units. The size limits refer to the smallest dimension of plates, prisms, and columns. If the units are more than twice the minimum size of "very coarse," the actual size is given: "prisms 30 to 40 cm across."

Grade: Grade describes the distinctness of units. Criteria are the ease of separation into discrete units and the proportion of units that hold together when the soil is handled. Three classes are used:

Weak: The units are barely observable in place. When gently disturbed, the soil material parts into a mixture of whole and broken units and much material that exhibits no planes of weakness. Faces that indicate persistence through wet-dry-wet cycles are evident if the soil is handled carefully. Distinguishing structurelessness from weak structure is sometimes difficult. Weakly expressed structural units in virtually all soil materials have surfaces that differ in some way from the interiors.

Moderate: The units are well formed and evident in undisturbed soil. When disturbed, the soil material parts into a mixture of mostly whole units, some broken units, and material that is not in units. Peds part from adjoining peds to reveal nearly entire faces that have properties distinct from those of fractured surfaces.

Strong: The units are distinct in undisturbed soil. They separate cleanly when the soil is disturbed. When removed, the soil material separates mainly into whole units. Peds have distinctive surface properties. The distinctness of individual structural units and the relationship of cohesion within units to adhesion between units determine grade of structure. Cohesion alone is not specified. For example, individual structural units in a sandy loam A horizon may have strong structure, yet they may be less durable than individual units in a silty clay loam B horizon of weak structure.

The degree of disturbance required to determine structure grade depends largely on moisture content and percentage and kind of clay. Only slight disturbance may be necessary to separate the units of a moist sandy loam having strong granular structure, while considerable disturbance may be required to separate units of a moist clay loam having strong blocky structure.

The three terms for soil structure are combined in the order (1) grade, (2) size, (3) shape. "Strong fine granular structure" is used to describe a soil that separates almost entirely into discrete units that are loosely packed, roughly spherical, and mostly between 1 and 2 mm in diameter.

The designation of structure by grade, size, and shape can be modified with other appropriate terms when necessary to describe other characteristics. Surface characteristics of units are described separately. Special structural units, such as the wedge-shaped units of Vertisols, are described in appropriate terms.

Compound Structure

Smaller structural units may be held together to form larger units. Grade, size, and shape are given for both and the relationship of one set to the other is indicated: "strong medium blocks within moderate coarse prisms," or "moderate coarse prismatic structure parting to strong medium blocky."

Extra-Structural Cracks

Cracks are macroscopic vertical planar voids with a width much smaller than length and depth. A crack represents the release of strain that is a consequence of drying. In many soils, cracks bound individual structural units. These cracks are repetitive and usually quite narrow. Their presence is part of the concept of the structure. The cracks to be discussed are the result of localized stress release which forms planar voids that are wider than the repetitive planar voids between structural units or which occur in massive or weakly structured material at relatively wide intervals. These cracks may be coextensive with crack space between structural units. If they are coextensive, the width exceeds that of the associated structural cracks. The areal percentage of such cracks, either on a vertical exposure or on the ground surface, may be measured by line-intercept methods. For taxonomic purposes, the width and depth of cracks has importance. Four kinds of extra-structural cracks may be recognized:

Surface-initiated reversible cracks form as a result of drying from the surface downward. They close after relatively slight surficial wetting and have little influence on ponded infiltration rates.

Surface-initiated irreversible cracks form on near-surface water reduction from exceptionally high water content related to freeze-thaw action and other processes. The cracks do not close completely when rewet and extend through the crust formed by frost action. They act to increase ponded infiltration rates.

Subsurface-initiated reversible cracks form as a result of appreciable reduction in water content from "field capacity" in horizons or layers with considerable extensibility. They close in a matter of days if the horizon is brought to the moderately moist or wetter state. They extend upward to the soil surface unless there is a relatively thick overlying horizon that is very weakly compacted (loose or very friable) and does not permit the propagation of cracks (mechanically bulked subzones). Such cracks importantly influence ponded infiltration rates and evaporation directly from the soil.

Subsurface-initiated irreversible cracks are the "permanent" cracks of the USDA soil taxonomy system. They have a similar origin to surface-initiated irreversible cracks, although quite different agencies are involved.

The foregoing genetic definition of cracks does not directly relate to prediction of infiltration. For such predictions, the surface connectiveness of the cracks and their depth must be specified. *Surface-connected cracks* occur at the ground surface or are covered by up to 10-15 cm of soil material that would permit the accumulation of free water at the plane that marks the top of the crack under conditions that may occur in most years.

If the antecedent water state of the overlying zone were very moist, free water from 25 mm of rainfall in one hour should reach the top of the cracks. Usually the zone would have *very high* or *high* saturated hydraulic conductivity. Such subzones may exhibit structure or be single grain. The structure units range widely in size. The common characteristic is that the consistent units of the mass as a whole are highly discrete and the porosity of the interstices among the structural units is high. If not too thick, the *mechanically bulked* subzone of tilled surface horizons would be such a zone.

A crack depth index value may be obtained by insertion of a blunt wire, approximately 2 mm in diameter. *Penetrant cracks* are 15 cm or more in depth as measured by wire insertion. Cracks that are both penetrant and surface connected are described as *penetrant surface connected*. Penetrant surface-connected cracks act to increase transient ponded infiltration. Prominence of the penetrant surface-connected cracks would depend on the linear distance of such cracks per unit area of ground surface. The linear distance may be allowed to decrease as crack depths increase. No classes are provided.

Internal Surface Features

Surface features include (1) coats of a variety of substances unlike the adjacent soil material and covering part or all of surfaces, (2) material concentrated on surfaces by the removal of other material, and (3) stress formations in which thin layers at the surfaces have undergone reorientation or packing by stress or shear. All differ from the adjacent material in composition, orientation, or packing.

Descriptions of surface features may include kind, location, amount, continuity, distinctness, and thickness of the features. In addition, colour, texture, and other characteristics that apply may be described, especially if they contrast with the characteristics of the adjacent material.

Kinds: Surface features are distinguished by differences in texture, colour, packing, orientation of particles, or reaction to various tests. If a feature is distinctly different from the adjacent material but kind cannot be determined, it is still described.

Clay films (synonymous with clay skins) are thin layers of oriented, translocated clay.

Clay bridges link together adjacent mineral grains.

Sand or silt coats are sand or silt grains adhering to a surface. Some sand and silt coats are concentrations of the sand and silt originally in the horizon from which finer particles have been removed. Some sand and silt coats are material that has been moved from horizons above and deposited on surfaces. In some coats the grains are almost free of finer material; in others, the grains themselves are coated. If known, the composition of the coat is noted.

Other coats are described by properties that can be observed in the field. The coats are composed variously of iron, aluminium or manganese oxides, organic matter, salts, or carbonates. Laboratory analyses may be needed for a positive identification.

Stress surfaces (pressure faces) are smoothed or smeared surfaces. They are formed through rearrangement as a result of shear forces. They may persist through successive drying and wetting cycles.

Slickensides are stress surfaces that are polished and striated and usually have dimensions exceeding 5 cm. They are produced by relatively large volumes of soil sliding over another. They are common below 50 cm in swelling clays which are subject to large changes in water state.

Location: The various surface features may be on some or all structural units, channels, pores, primary particles or grains, soil fragments, rock fragments, nodules, or concretions. The kind and orientation of surface on which features are observed is always given. For example, if clay films are on vertical but not horizontal faces of peds, this fact should be recorded.

Amount: The percentage of the total surface area of the kind of surface considered occupied by a particular surface feature over the extent of the horizon or layer is described. Amount can be characterized by the following classes:

very few: Occupies < 5 percent.

few: Occupies 5 to 25 percent.

common: Occupies 25 to 50 percent.

many: Occupies > 50 percent.

The same classes are used to describe the amount of "bridges" connecting particles. The amount is judged on the basis of the percentage of particles of the size designated that are joined to adjacent particles of similar size by bridges at contact points.

Distinctness: Distinctness refers to the ease and degree of certainty with which a surface feature can be identified. Distinctness is related to thickness, colour contrast with the adjacent material, and other properties. It is, however, not itself a measure of any one of them. Some thick coats, for example, are faint; some thin ones are prominent. The distinctness of some surface features changes markedly as water state changes. Three classes are used.

Faint: Evident only on close examination with 10X magnification and cannot be identified positively in all places without greater magnification. The contrast with the adjacent material in colour, texture, and other properties is small.

Distinct: Can be detected without magnification, although magnification or tests may be needed for positive identification. The feature contrasts enough with the adjacent material to make a difference in colour, texture, or other properties evident.

Prominent: Conspicuous without magnification when compared with a surface broken through the soil. Colour, texture, or some other property or combination of properties contrasts sharply with properties of the adjacent material or the feature is thick enough to be conspicuous.

The order of description is usually amount, distinctness, colour, texture, kind, and location. Two examples: "few distinct grayish brown (10YR 5/2) clay films on vertical faces of peds"; "many distinct brown clay bridges between mineral grains." Only properties are listed that add to the understanding of the soil. If texture of the surface feature is obvious, as in most stress surfaces, repeating texture adds nothing. Kind and location are essential if the feature is mentioned at all. The conventions do not characterize the volume of the surface features. If volume is important, it is estimated separately.

Concentrations

The features discussed here are identifiable bodies within the soil that were formed by pedogenesis. Some of these bodies are thin and sheet-like; some are nearly equidimensional; others have irregular shapes. They may contrast sharply with the surrounding material in strength, composition, or internal organization. Alternatively, the differences from the surrounding material may be slight. Soft rock fragments which have rock structure but are *weakly cemented* or *noncemented* are not considered concentrations. They are excluded on the basis of inference as to a geological as opposed to a pedological origin.

Masses are noncemented concentrations of substances that commonly cannot be removed from the soil as a discrete unit. Most accumulations consist of calcium carbonate, fine crystals of gypsum or more soluble salts, or iron and manganese oxides. Except for very unusual conditions, masses have formed in place.

Plinthite consists of reddish, iron-enriched bodies that are low in organic matter and are coherent enough to be separated readily from the surrounding soil. Plinthite commonly occurs within and above reticulately mottled horizons. Plinthite has higher penetration resistance than adjacent brown or gray bodies or than red bodies that do not harden. Soil layers that contain plinthite rarely become dry in the natural setting.

The bodies are commonly about 5 to 20 mm across their smallest dimension. Plinthite bodies are *firm* or *very firm* when moist, *hard* or *very hard* when air dry, and become *moderately cemented* on repetitive wetting and drying. They occur as discrete nodules or plates. The plates are oriented horizontally. The nodules occur above and the plates within the upper part of the reticulately mottled horizon. The plates generally have a uniformly reddish colour and have sharp boundaries with the surrounding brown or gray material. The part of the iron-rich body that is not plinthite normally stains the fingers when rubbed while wet, but the plinthite centre does not. It has a harsh, dry feel when rubbed, even if wet. Horizons containing plinthite are more difficult to penetrate with an auger than adjacent horizons at the same water state and clay content but which do not contain plinthite. Plinthite generally becomes less cemented after prolonged submergence in water. An air dry sample can be dispersed by normal procedures for particle-size distribution.

Nodules and *concretions* are cemented bodies that can be removed from the soil intact. Composition ranges from material dominantly like that of the surrounding soil to nearly pure chemical substances entirely different from the surrounding material. Their form is apparently not governed by crystal forms based on examination at a magnification of 10X as is the case for crystals and clusters of crystals. It is impossible to be sure if some certain nodules and concretions formed where they are observed or were transported.

Concretions are distinguished from nodules on the basis of internal organization. Concretions have crude internal symmetry organized around a point, a line, or a plane. Nodules lack evident, orderly, internal organization. A typical example of a concretion organized around a point. The internal structure typically takes the form of concentric layers that are clearly visible to the naked eye. A coat or a very thin outer layer of an otherwise undifferentiated body does not indicate a concretion.

Crystals are considered to have been formed in place. They may occur singly or in clusters. Crystals of gypsum, calcite, halite, and other pure compounds are common in some soils. These are described as crystals or clusters of crystals, and their composition is given if known.

Ironstone is an in-place concentration of iron oxides that is at least weakly cemented. Ironstone nodules are commonly found in layers above plinthite. These ironstone nodules are apparently plinthite that has cemented irreversibly as a result of repeated wetting and drying. Commonly, the centre of iron-rich bodies cements upon repeated wetting and drying but the periphery does not.

Describing Concentrations Within the Soil

Any of a large number of attributes of concentrations within the soil may be important; the most common are number or amount, size, shape, consistence, colour composition, kind, and location. Not all of these attributes are necessarily described. The order as listed above is convenient for describing them, as for example: "many, fine, irregular, hard, light gray, carbonate nodules distributed uniformly through the horizon." The conventions for describing kind have been indicated in this section. Descriptions of consistence and colour are discussed in other parts of this chapter.

Amount or *quantity* of concentrations refers to the relative volume of a horizon or other specified unit occupied by the bodies. The classes used for quantity of mottles are also used for these features.

Size may be measured directly or given by the classes listed below. The dimension to which size-class limits apply depends on the shape of the body described. If the body is nearly uniform, size is measured in the shortest dimension, such as the effective diameter of a cylinder or the thickness of a plate. For irregular bodies, size refers to the longest dimension unless that creates an erroneous impression; measurements can be given if needed.

The following size classes are used:

- *fine:* < 2 mm
- *medium:* 2- 5 mm
- *coarse:* 5- 20 mm

- *very coarse:* 20- 76 mm
- *extremely coarse:* > 76 mm.

Shape of concentrations is variable both among kinds of concentrations and commonly within a concentration. The following terms are suggested:

Rounded: Approximately equidimensional, a few sharp corners, and at least approximately regular.

Cylindrical: At least crudely cylindrical or tubular; one dimension is much greater than the other two.

Platelike: Shaped crudely like a plate; one dimension is very much smaller than the other two. The term "platelike" is used to avoid confusion with platy structure.

Irregular: Characterized by branching, convoluted, or mycelial form.

The terms listed apply to all concentrations. Individual crystals of a particular mineral usually implies a shape.

Composition of bodies is described if known and if important for understanding their nature or the nature of the soil in which they are observed. Some of the physical attributes of the interior of a feature are implied by the name. Other features, such as enclosed mineral grains, patterns of voids, or similarity to the surrounding soil, may be important. A distinction is made between bodies that are composed dominantly of a single substance and those that are composed of earthy material impregnated by various substances. For many bodies, the chemical composition cannot be determined with certainty in the field. The following set of terms, however, is useful for describing composition. If the substance dominates the body, then the body is described as a substance body. If the substance impregnates other material, the body is described as a body of substance accumulation.

Carbonates and iron are common substances that dominate or impregnate nodular or concretionary bodies. Discrete nodules of clay are found in some soils; argillaceous impregnations are less common. Materials dominated by manganese are rare, but manganese is conspicuous in some nodules that are high in iron and mistakenly called "manganese nodules."

Consistence

Soil consistence in the general sense refers to "attributes of soil material as expressed in degree of cohesion and adhesion or in resistance to deformation on rupture." As employed here consistence includes: (1) resistance of soil material to rupture, (2) resistance to penetration, (3) plasticity, toughness, and stickiness of puddled soil material, and (4) the manner in which the soil material behaves when subject to compression. Although several tests are described, only those should be applied which may be useful.

A word may be in order about the similar term, consistency. Consistency was used originally in soil engineering for a set of classes of resistance to penetration by thumb or thumbnail. The term has been generalized to cover about the same concept as "consistence." The set of tests specified, however, is different from those given here.

Consistence is highly dependent on the soil-water state and the description has little meaning unless the water state class is specified or is implied by the test. Previously class sets were given for "dry" and "moist" consistence of the soil material as observed in the field. "Wet" consistence was evaluated for puddled soil material. Here the terms used for "moist" consistence previously are applied to the wet state as well. The previous term "wet consistence" is dropped. Stickiness, plasticity, and toughness of the puddled soil material are independent tests.

For determinations on the natural fabric, variability among specimens is likely to be large. Multiple measurements may be necessary. Recording of median values is suggested in order to reduce the influence of the extremes measured.

Rupture Resistance Block-like Specimens

The classes of resistance to rupture and the means of determination for specimens that are block-like. Different class sets are provided for moderately dry and very dry soil material, and for slightly dry and wetter soil material. Unless specified otherwise, the soil-water state is assumed to be that indicated for the horizon or layer when described. Cementation is an exception. To test for cementation, the specimen is air-dried and then submerged in water for at least 1 hour. The placements do not pertain to the soil material at the field water state. The blocklike specimen should be 25-30 mm on edge. Direction of stress relative to the in-place axis of the specimen is not defined unless otherwise indicated. The specimen is compressed between extended thumb and forefinger, between both hands, or between the foot and a nonresilient flat surface. If the specimen resists rupture by compression, a weight is dropped onto it from increasingly greater heights until rupture.

Failure is at the initial detection of deformation or rupture. Stress applied in the hand should be over a 1-second period. The tactile sense of the class limits may be learned by applying force to top loading scales and sensing the pressure through the tips of the fingers or through the ball of the foot. Postal scales may be used for the resistance range that is testable with the fingers. A bathroom scale may be used for the higher rupture resistance. Specimens of standard size and shape are not always available. Blocks of specimens that are smaller than 25-30 mm on edge may be tested. The force withstood may be assumed to decrease as the reciprocal of the dimension along which the stress is applied. If a block specimen with a length of 10 mm along the direction the force is applied were to be ruptured, the force should be one-third that for an identical specimen 30 mm on edge. If the specimen is smaller than the standard size, the evaluated rupture resistance should be recorded and the dimensions of the specimen along the axis the stress is applied should be indicated.

Soil structure complicates the evaluation of rupture resistance. If a specimen of standard size can be obtained, report the rupture resistance of the standard specimen and other individual constituent structural units as desired. Usually the constituent structural units must exceed about 5 mm in the direction the stress is applied; expression must exceed weak for the rupture resistance to be evaluated. If structure size and expression are such that a specimen of standard size cannot be obtained, then the soil material overall is loose. Structural unit resistance to rupture may be determined if the size is large enough (exceed about 5 mm in the direction stress is applied) for a test to be performed.

Rupture Resistance Plate-Shaped Specimens

Tests are described that are applicable to plate-shaped specimens where the length and width are several times more than the thickness. Test procedures were developed for surface crusts but are applicable to plate-shaped bodies at greater depth in the soil. An alternative method of directly measuring plate-shaped specimens is to break them into a crudely blocked form. If the dimensions of the resulting block specimens are smaller than 25-30 mm on edge, it would be assumed that the measured rupture resistance is lower by 25.

Rupture Resistance by Crushing: This test was designed primarily for air dry surface crust, but it may be used for other soil features. The morphological description of surface crust is discussed earlier in this chapter. The specimen should be 10 to 15 mm on edge and 5 mm thick or the thickness of occurrence if less than 5 mm. If surface crust, the thickness is inclusive of the crust proper and the adhering soil material beneath. The specimens are small to make the test applicable to crusts with closely spaced cracks. The specimen is grasped on edge between extended thumb and first finger. Force is applied along the longer of the two principal dimensions. Compression to failure should be over about one second. A scale may be used to both rupture the specimens directly and develop the finger tactile sense. Force is applied with the first finger through a bar 5 mm across on the scale to create a similar bearing area to that of the plate-like specimen. The specimen is compressed between thumb and first finger while simultaneously exerting the same felt pressure on the scale with the first finger of the other hand.

The scale is read at the failure of the specimen. For specimens that cannot be broken between thumb and forefinger, the resistance to rupture may be evaluated using a small penetrometer. The specimen is formed with the two larger surfaces parallel and flat. The specimen is placed with a larger face downward on a nonresilient surface and force is applied through the 6 mm diameter penetrometer tip until rupture occurs.

For plate-shaped bodies that are durable enough to withstand handling, such as fragments of fissile sedimentary rock, a modulus of rupture estimation is an appropriate test. In practice, modulus of rupture tests commonly would be used to acquire a tactile sense which then would be used directly in the field. Insufficient experience has been obtained to provide classes. The tests to follow are hand-held tests. The configuration of the tests do not conform rigorously to the requirements for measurement of modulus of rupture. Furthermore, the amount of force applied may be only roughly approximated. For these reasons, the test results are only a crude measure of the modulus of rupture. In one test, a specimen is held in contact with a small diameter cylindrical shaft (pencil, nail, and so on) placed near the centre of the specimen. Stress is applied by pressing in opposite directions with the two first fingers and the thumbs until rupture occurs. The equation for the modulus of rupture in MPa is:

$$S = (0.15FL)/(bd^2)$$

where F is the force in newtons, L is the distance between the shaft and the inside edge of the area over which the force is applied on either side of the shaft with the fingers, b is the width of the specimen (in centimetres), and d is the depth or thickness in the direction of the load (in centimetres). The force application is based on the tactile sense and hence is approximate.

In the other approach, the specimen is grasped firmly at one end with pliers and force is applied downward at an established distance (to the nearest 1 cm) from the edge of the pliers. The area over which the force is applied should be small. The flat-end rod penetrometer described in the section on micropenetration resistance works well. A chisel point may be mounted over the tip. Modulus of rupture, S, expressed in megapascals (MPa), is calculated by:

$$S = (0.6FL)/(bd^2)$$

where F is the force in newtons, L is the distance between the end of the jaws of the pliers and the inside edge of the area where the force is applied, b is width, and d is the thickness. The dimensions are all in centimetres. Length and width are estimated to 1 cm and thickness to 1 mm.

Plasticity

Plasticity is the degree to which puddled soil material is permanently deformed without rupturing by force applied continuously in any direction. Plasticity is determined on material smaller than 2 mm. The determination is made on thoroughly puddled soil material at a water content where maximum plasticity is expressed. This water content is above the plastic limit, but it is less than the water content at which maximum stickiness is expressed. The water content is adjusted by adding water or removing it during hand manipulation. The closely related plastic limit that is used in engineering classifications is the water content for < 0.4 mm material at which a roll of 3 mm in diameter which had been formed at a higher water content breaks apart.

Animals

Mixing, changing, and moving of soil material by animals is a major factor affecting properties of some soils. The features left by the work of some animals reflect mainly mixing or transport of material from one part of the soil to another or to the surface. The original material may be substantially modified physically or chemically.

Cicada casts at about 0.4 actual size. The photograph is a close-up view of an indurated horizon about 35 cm thick. Cicada casts in varying stages of induration are common in some soils of semi-arid climates. The features that animals produce on the land surface may be described. Termite mounds, ant hills, heaps of excavated earth beside burrows, the openings of burrows, paths, feeding grounds, earthworm or other castings, and other traces on the surface are easily observed and described.

Simple measurements and estimates—such as the number of structures per unit area, proportionate area occupied, volume of above-ground structures—give quantitative values that can be used to calculate the extent of activity and even the number of organisms.

The marks of animals below the ground surface are more difficult to observe and measure. Observations are confined mainly to places where pits are dug. The volume of soil generally studied is limiting. For the marks of many animals, the normal pedon for soil characterization is large enough to provide a valid estimate. For some animals, however, the size of the marks is too large for the usual pedon. The features produced by animals in the soil are described in terms of amount, location, size, shape, and arrangement, and also in terms of the colour, texture, composition, and other properties of the component material. No special conventions are provided. Common words should be used in conjunction with appropriate special terms for the soil properties and morphological features that are described elsewhere in this manual.

Krotovinas are irregular tubular streaks within one layer of material transported from another layer. They are caused by the filling of tunnels made by burrowing animals in one layer with material from outside the layer. In a profile, they appear as rounded or elliptical volumes of various sizes. They may have a light colour in dark layers or a dark colour in light layers, and their other qualities of texture and structure may be unlike those of the soil around them.

Selected Chemical Properties

This section discusses selected chemical properties that are important for describing and identifying soils.

Reaction: The numerical designation of reaction is expressed as pH. With this notation, pH 7 is neutral. Values lower than 7 indicate acidity; values higher, indicate alkalinity. Most soils range in pH from slightly less than 2.0 to slightly more than 11.0, although sulfuric acid forms and pH may decrease to below 2.0 when some naturally wet soils that contain sulfides are drained.

The descriptive terms to use for ranges in pH are as follows:

Term	Range
Ultra acid	< 3.5
Extremely acid	3.5 - 4.4
Very strongly acid	4.5 - 5.0

Strongly acid	5.1 - 5.5
Moderately acid	5.6 - 6.0
Slightly acid	6.1 - 6.5
Neutral	6.6 - 7.3
Slightly alkaline	7.4 - 7.8
Moderately alkaline	7.9 - 8.4
Strongly alkaline	8.5 - 9.0
Very strongly alkaline	> 9.0

Both colorimetric and electrometric methods are used for measuring pH. Colorimetric methods are simple and inexpensive. Reliable portable pH meters are available.

Carbonates of Divalent Cations

Cold 2.87N (about a 1:10 dilution of concentrated HCl) hydrochloric acid is used to test for carbonates in the field. The amount and expression of effervescence is affected by size distribution and mineralogy as well as the amount of carbonates. Consequently, effervescence cannot be used to estimate the amount of carbonate. Four classes of effervescence are used:

Very slightly effervescent: few bubbles seen

Slightly effervescent: bubbles readily seen

Strongly effervescent: bubbles form low foam

Violently effervescent: thick foam forms quickly

Calcium carbonate effervesces when treated with cold dilute hydrochloric acid. Effervescence is not always observable for sandy soils. Dolomite reacts to cold dilute acid slightly or not at all and may be overlooked.

Dolomite can be detected by heating the sample, by using more concentrated acid, and by grinding the sample. The effervescence of powdered dolomite with cold dilute acid is slow and frothy and the sample must be allowed to react for a few minutes.

Salinity and Sodicity

Accurate determinations of salinity and sodicity in the field require special equipment and are not necessarily part of each pedon investigation. Reasonable estimates of salinity and sodicity can be made if field criteria are correlated to more precise laboratory measurement.

Salinity

The electrical conductivity of a saturation extract method is the standard measure of salinity. Electrical conductivity is related to the amount of salts more soluble than gypsum in the soil, but it may include a small contribution (up to 2 dS/m) from dissolved gypsum.

The standard international unit of measure is decisiemens per meter (dS/m) corrected to a temperature of 25 °C. Millimhos per centimetre (mmhos/cm) means the same as dS/m and may still be used. If it has been measured, the electrical conductivity is reported in soil descriptions. The following classes of salinity are used if the electrical conductivity has not been determined, but salinity is inferred:

	Class	Electrical conductivity
		dS/m (mmhos/cm)
0	Non saline	0 - 2
1	Very slightly saline	2 - 4
2	Slightly saline	4 - 8
3	Moderately saline	8 - 16
4	Strongly saline	> 16

Sodicity

The sodium adsorption ratio (SAR) is the standard measure of the sodicity of a soil. The sodium adsorption ratio is calculated from the concentrations (in milliequivalents per litter) of sodium, calcium, and magnesium in the saturation extract:

$$SAR = \frac{Na^+}{\sqrt{\dfrac{Ca^{++} + Mg^{++}}{2}}}$$

Formerly, the exchangeable sodium percentage, which equals exchangeable sodium (meq/100 g soil) divided by the cation exchange capacity (meq/100 g soil) times 100, was the primary measure of sodicity. The test for exchangeable sodium percentage, however, has proved unreliable in soils containing soluble sodium silicate minerals or large amounts of sodium chloride. Sodium is toxic to some crops, and sodium affects the soil's physical properties, mainly saturated hydraulic conductivity. A sodic condition has little effect on hydraulic conductivity in highly saline soils. A soil that is both saline and sodic may, when artificially drained, drain freely at first. After some of the salt has been removed, however, further leaching of salt becomes difficult or impossible. The sodium adsorption ratio (SAR) usually decreases as a soil is leached, but the amount of change depends in part on the composition of the water used for leaching and, therefore, cannot be predicted with certainty. If the initial SAR is greater than 10 and the initial electrical conductivity is more than 20 dS/m and information is needed as to whether the soil will be sodic following leaching, the SAR is determined on another sample after first leaching with the intended irrigation water. For the land reclamation of soils with an electrical conductivity of more than 20 dS/m, the SAR is used that is determined after leaching with distilled water to an electrical conductivity of about 4 dS/m.

Sulfates: Gypsum (calcium sulfate) can be inherited from the parent material, or it can precipitate from supersaturated solutions in the soil or in the substratum. Gypsum can alleviate the effects of sodium, making possible the use of irrigation water that has a relatively high amount of sodium. Soils that contain large amounts of gypsum can settle unevenly after irrigation; frequent releveling may be required. Gypsum is soluble in water. The electrical conductivity of a distilled water solution with gypsum is about 2dS/m. In the absence of other salts, a salinity hazard does not exist except for such sensitive plants as strawberries and some ornamentals. Gypsum and other sulfates may cause damage to concrete.

Much gypsum is tabular or fibrous and tends to accumulate as clusters of crystals or as coats on peds. Some of it is cemented. Gypsum can usually be identified tentatively by its form and lack of effervescence with acid. Gypsum in the parent material may not be readily identifiable. If determined, the amount of gypsum is shown in the description; otherwise, the amount may be estimated. Semiquantitative field methods for determining amounts of gypsum are available. A few soils contain large amounts of sodium sulfate, which looks like gypsum. At temperatures above 32.4 °C it is in the form of thenardite (Na_2SO_4) and at lower temperatures in the form of mirabilite ($Na_2SO_4 \cdot 10H_2O$). The increase in volume and decrease in solubility as thenardite changes to mirabilite can cause spectacular salt heaving. In sodium-affected soils, sodium sulfate is a common water-soluble salt.

Bibliography

Adhikari M.K. : *Research on Plant Tissue Culture*, Department of Plant Resources National Herbarium and Plant Laboratories, 2004.

Alfred Steferud: *Diseases of Fruits and Nuts*, Biotech Books, Delhi, 2005.

Alka Rani Upadhyay: *Aquatic Plants for the Waste Water Treatment*, Daya, Delhi, 2004.

Amarnath, J S and A P V Samvel: *Agri-Business Management*, Satish Serial Pub, Delhi, 2008.

Ansari, Tariq M : *Molecular Plant Pathology*, Pearl Books, Delhi, 2008.

Asaithambi S. : *Economics of Ground Water Management in India*, Abhijeet, Delhi, 2008.

Ashworth S.: *Seed to Seed*, Decorah, Seed Savers Publications, 1991.

Asit K. Biswas and Cecilia Tortajada: *Appraising Sustainable Development: Water Management and Environmental Challenges*, Oxford University Press, Delhi, 2005.

Bagis, Ali Ihsan: *Water as an Element of Cooperation and Development in the Middle East*, Ankara, Ayna Publications, 1994.

Bahar A. Siddiqui and Samiullah Khan: *Plant Breeding Advances and in vitro Culture*, CBS, Delhi, 1997.

Bandyopadhyay, P C : *Breeding and Crop Production*, Gene Tech Books, Delhi, 2007.

Bhakar S.R. : *Ground Water Hydrology : Theory and Practice*, Agrotech, Delhi, 2009.

Bhave, P.R. and R. Gupta: *Analysis of Water Distribution Networks*, Narosa, Delhi, 2011.

Biswas Asit K. : *Integrated Water Resources Management in South and South-East Asia*, , Oxford University Press, Delhi, 2001.

Bourne, Peter G.: *Water and Sanitation: Economic and Sociological Perspectives* Academic Press. Orlando. 1984.

Boyer , J.S.: *Measuring the Water Status of Plants and Soils*, Academic Press, N.Y., 1995.

Brooks, D.: *Water: Local-Level Management*, Ottawa, International Development Research Centre, 2002.

Brown R G : *Dictionary of Plant Tissue Culture*, Ivy Pub, Delhi, 2004.

Byrd, Graf: *Advances in Plant Physiology*, Rajat Pub, Delhi, 2008.

Chadha K. L. and Pareek O. P.: *Advances in Horticulture: Fruit Crops,* New Delhi, Malhotra Publishing House, 1993.

Chahal, S.S. : *Achievements and Prospects in Mycology and Plant Pathology*, International, Delhi, 1997.

Chambers, R.: *Rural Development: Putting the Last First*, London, Longman, 1983.

Chandola, R.P. : *Dictionary of Plant Pathology*, Daya, Delhi, 2010.

Collymore L.: *Fruit Production in Barbados*, Port of Spain, Trinidad and Tobago, 1996.

Cullis, A. : *Rainwater Harvesting: The Collection of Rainfall and Runoff in Rural Areas,* London, U.K.: IT Publications, 1986.

Dabholkar, A.R. : *General Plant Breeding*, Concept, Delhi, 2006.

Devi, C.R. Sudharmai : *Analytical Procedures in Soil Science and Agricultural Chemistry*, Agrotech, Delhi, 2004.

Doijode S. D.: *Seed Germination in Fruits*, New Delhi, Malhotra Publishers, 1993.

Easter, K. William : : *Irrigation Investment Technology, and Management Strategies for Development Studies in Water Policy Management 9*, Boulder, Colo Westview Press, 1986.

Engelman, R., and P. LeRoy : *Sustaining Water Washington*, D.C.: Population Services International, New York, 1993.

Erik Nissen-Peterson : *Rainwater Catchment Systems*, UK: Intermediate Technology Publications, 1999.

Ferentinos L.: *Proceeding of the Sustainable Taro Culture for the Pacific Conference*, Honolulu, HITAHR, 1993.

Ghosh, N.C. and K.D. Sharma: *Groundwater Modelling and Management*, Capital Pub, Delhi, 2006.

Gould, J., and E. Nissen-Petersen : *Rainwater Catchment Systems for Domestic Supply: Design, Construction and Implementation,* London, U.K.: IT Publications, 1999.

Gour, H.N.: *Integrated Plant Pathology*, Scientific, Delhi, 2009.

Graf, Alfred Byrd : *Advances in Plant Physiology*, Rajat Pub, Delhi, 2008.

Gupta, O.P. : *Water in Relation to Soils and Plants : With Special Reference to Agriculture*, Agrobios, Delhi, 2002.

Gurjar, Ram Kumar : *Geography of Water Resources*, Rawat, Delhi, 2008.

Helen Bannayan: *Water Resources of Jordan: Present Status and Future Potentials*, Amman, Friedrich-Ebert-Stiftung, 1993.

Herminie Broedel Kitchen: *Soils and Crops : Diagnostic Techniques*, Satish Serial Publishing, Allahabad, 2004.

Husain, Ahmad : *Environment and Water Resource Management*, Sumit Enterprises, Delhi, 2006.

Jacquat Christiane: *Plants from the Markets of Thailand*, Bangkok, Duang Kamol, 1990.

Jana B L : *Water Harvesting and Watershed Management*, Agrotech, Delhi, 2008.

Jeet Inder : *Rainwater Harvesting*, Mittal, Delhi, 2009.

Jha Timir Baran and Ghosh Biswajit : *Plant Tissue Culture : Basic and Applied*, Universities Press, Delhi, 2005.

Jones, R. M.: *Plant Resources of South-East Asia,* Wageningen, Pudoc Scientific Publishers, 1992.

Kanmony, J. Cyril : *Drinking Water Management : Problems and Prospects*, Mittal Pub, Delhi, 2010.

Kapoor, R.L. and M.L. Saini: *Plant Breeding and Crop Improvement*, CBS, Delhi, 1997.

Kataria, T N : *Plant and Crop Physiology*, Pearl Books, Delhi, 2008.

Khan M A : *Water Resources Management and Sustainable Agriculture*, APH, Delhi, 2008.

Kumar Shailesh : *Plant Tissue Culture Theory and Techniques*, Scientific, Delhi, 2009.

Kumar U. : *Methods in Plant Tissue Culture*, Agrobios, Delhi, 2003.

Kumar, N. : *Breeding of Horticultural Crops : Principles and Practices*, New India Pub, Delhi, 2006.

LeRoy, K. : *Sustaining Water Washington*, D.C.: Population Services International, New York, 1993.

Linda M. Welkom: *Groundwater Chemicals Desk Reference*, Chelsea, MI, 1990.

Macself, R : *Soils and Fertilizers*, Satish Serial Pub, Delhi, 2005.

Mahajan Gautam : *Ground Water : Surveys and Investigation*, APH, Delhi, 2009.

Malins A.: *Postharvest Handling of Pineapple and Mango*, Port of Spain, Trinidad and Tobago, 1992.

McGarry, M.G.: *Matching Water Supply Technology to the Needs and Resources of Developing Countries,* United Nations. New York. 1987

Meenu Bhatnagar: *Groundwater Management : Sustainable Approaches*, Icfai Books, Delhi, 2012.

Mitra S.: *Postharvest Physiology and Storage of Tropical and Subtropical Fruits*, Oxon, CABI, 1997.

Murakami, Masahiro, *Managing Water for Peace in the Middle East : Alternative Strategies*, Tokyo and NYC, United Nations University Press, 1995.

Narasaiah, M. Lakshmi : *Energy, Irrigation and Water Supply*, Discovery, Delhi, 2004.

Nielsen, David M.: *Practical Handbook of Groundwater Monitoring*. Lewis Publishers, Chelsea, 1991.

Patel, GM: *Botanical Pesticides for Pest Management*, Scientific, Delhi, 2008.

Pilgrim R.: *Post Harvest Handling of Minor Exotics*, St. George's, Grenada, 1996.

Premjit Sharma: *Agricultural Drainage and Water Quality*, Gene Tech Books, Delhi, 2007.

Rao K. Nageswara : *Water Resources Management : Realities and Challenges*, New Century Publication, Delhi, 2006.

Sambamurty, A V S S : *Textbook of Plant Pathology*, I K Pub, Delhi, 2006.

Santhi, R and K M Sellamuthu: *Fundamentals of Forest Soils*, Satish Serial Pub, Delhi, 2008.

Sathyanarayana B N : *Plant Tissue Culture : Practices and New Experimental Protocols*, I K International, Delhi, 2007.

Sharma, S.S.P. and U.H. Kumar: *Dynamics of Watershed Development and Livelihood in India*, Serials Pub, Delhi, 2010.

Singh A K : *Environment and Water Resources Management*, Adhyayan, Delhi, 2006.

Singh, S K : *Biotechnology, Plant Propagation and Plant Breeding*, Campus Books, Delhi, 2008.

Somani, L L and P C Kanthaliya: *Soils and Fertilisers at a Glance*, Agrotech, Delhi, 2004.

Srivastava, Seema : *Plant Physiology and Biochemistry*, Campus Books, Delhi, 2009.

Thapliyal, B K : *Democratisation of Water*, Serials Pub, Delhi, 2008.

Thind, T.S. : *Annual Review of Plant Pathology*, Scientific, Delhi, 2004.

Thomas E.: *Fruit Production in St. Kitts and Nevis*, Port of Spain, IICA, 1996.

Tripathi, D P : *Plant Pathology at a Glance*, Scientific, Delhi, 2008.

Tyagi, I.D. : *Plant Breeding and Genetics at a Glance*, South Asian, Delhi, 2005.

Verma J.P. : *Detection of Plant Pathogens and their Management*, Angkor, Delhi, 1995.

Walton, William C.: *Groundwater Resource Evaluation*. McGraw Hill, New York, 1970.

Whealy K.: *The Garden Seed Inventory*, Decorah, Seed Saver Publications, 1988.

Index

ooo